The Sea Inside

By the same author

Serious Pleasures: The Life of Stephen Tennant

Noël Coward: A Biography

*Wilde's Last Stand: Decadence, Conspiracy and
the First World War*

Spike Island: The Memory of a Military Hospital

England's Lost Eden: Adventures in a Victorian Utopia

Leviathan or, The Whale

The Sea Inside

PHILIP HOARE

FOURTH ESTATE·*London*

First published in Great Britain in 2013
by Fourth Estate

An imprint of HarperCollins*Publishers*
77–85 Fulham Palace Road,
Hammersmith, London W6 8JB
www.4thestate.co.uk

A catalogue record for this book is available from
the British Library

ISBN 978 0 00 741211 2

Printed and bound in Great Britain by Clays Ltd, St Ives plc

Typeset in Emona by Terence Caven

The Leverhulme Trust

MIX
Paper from
responsible sources
FSC™ C007454

FSC is a non-profit international organisation established to promote
the responsible management of the world's forests. Products carrying
the FSC label are independently certified to assure customers that
they come from forests that are managed to meet the social, economic
and ecological needs
of present and future generations.

Find out more about HarperCollins and the environment at
www.harpercollins.co.uk/green

For Cyrus, Max & Lilian

Contents

...Even now my heart
Journeys beyond its confines, and my thoughts
Over the sea, across the whale's domain,
Travel afar the regions of the earth...

'The Seafarer', Anglo-Saxon verse

1
The suburban sea

I have lived long enough in the Shire to be
able to afford to go away from it with pleasure.
I suppose this is what homes are for. If one
hadn't got an anchorage it wouldn't be
exciting to sail away.

T.H. WHITE, *England Have My Bones,* 1936

In the years since I have come back to it, the house has grown to become part of me. It is held together by memories, even as it is falling apart. Surrounded by ivy and screened by trees, it has become an enclosed world, left to itself. As I look out from my bedroom window, a blackbird paces out the garage roof, over which a broken willow hangs. Below, tadpoles swim blindly in a slowly leaking pool.

Every day here is the same. I work at the same time, eat my meals at the same time, go out at the same time, go to bed at the same time. I hold fast to my routine, anchoring a life that might otherwise come adrift. But at night the anarchy of my dreams disturbs this self-imposed regime, freefalling till I regain the rituals of the morning.

I'm woken by the cold outside my window, the dark pressing against the glass. I listen to the litany of the forecast, taking me around the coast –

Dogger, Fisher, German Bight, Humber, Thames,
Dover, Wight, Portland, Plymouth, Biscay, Trafalgar,
FitzRoy, Sole, Lundy, Fastnet, Irish Sea, Shannon,
Rockall, Malin, Hebrides, Bailey, Fair Isle, Faeroes

– places I'll never visit but whose names reassure me with their familiar rhythms, while their remote conditions seem strangely consoling.

Low, losing its identity, mainly fair, moderate or
good, falling more slowly, mainly fair in west, occa-
sionally poor in east, good.

I pay attention to the wind direction; not for my boat, but for my bike. A northerly means a chilly but fast ride south, an uphill struggle on my return. A prevailing sou'westerly signals a speedy cycle home, the wind behind my back like a sail. I peer through the curtains for a faint sign that the long night is drawing to a close.

Variable three or less, fog patches at first, becoming
mainly good.

Outside, I smell the night-morning air; promising and inviting, or closed-down and denying. I unlock the old brass padlock on the garage door and pull out my bike, feeling for its cable lock like reins.

I follow the same route. I've been doing it so long that my bike could steer itself to its destination. Its tyres know every crack in the tarmac, every worn-out white line, every pothole. The urgency of my mission leaves the sleeping houses behind. I ride like my mother walked, at a furious pace, always trying to keep up with myself. As though, if I went fast enough, no one would see me.

It's December. The colour has yet to seep back into the sky, before the warmth of the night meets the cold of the dawn. My body is geared to the bike. Green lights signal to non-existent cars; I sail through on red. Sodium lamps soak the streets in a Lucozade haze. I ride down the white lines, arms outstretched, reclaiming the road.

Soon I pass out of the suburb's sprawl, from the land to the sea. Crunching through the shingle, along furrows made on earlier visits, I rumble to a halt at the appointed spot. Leaning my bike against the sea wall, I climb over and – no going back now – lower myself in.

The water is so clear it scares me. Fish leap up as though they'd dropped out of the clouds. Everything is rising to the surface, summoned by the light, slowed to the sea's heartbeat. The water brims like an overrun bath. I push out through the stillness of the standing tide, my hands creating the only ripples.

I drift out as far as I dare. Borne up by the water, I turn briefly on my back, hips held to the sky, before striking back for the shore. I haul myself out, my body pink and steaming like a wet dog. The scar on my knee is a livid purple; my white T-shirt glows blue in the faint light.

Birds become visible and audible before the sun rises and the world wakes, a netherworld neither dark nor light, out of time. The anonymous Anglo-Saxon poet of 'The

Wanderer' had a word for it, *ūhta*, 'the hour before dawn', as he travelled by winter, watching 'the sea birds bathing, stretching out their wings'.

It's over all too quickly. A frail sun appears over the trees, milky-lemon pale, more like the moon. The morning begins. My body suffers but my spirit soars at having stolen a march on the day; having been rewarded as well as punished. I've forgotten my gloves. I stuff one hand, then the other, in my pockets. A skein of geese fly low over the mud. Curlew call out their names – *curl-you, curl-you* – whistling through their arched bills. Gulls clamour.

Cloud layers over cloud, plum, purple and navy, cocoa, khaki and grey. The cold surrounds me, almost comforting. The light lifts and falls. I smell the iodine of the shore and an intimation of drizzle in the air as the birds' noise rises in a crescendo, an orchestrated start to the day. The docks rumble into action. A brightly-lit liner sails up the estuary, silent yet so full of people, a distant glimpse of glamour.

The night fades away. I race the ship back to port, flushing a lone and indignant duck from the freshwater pond that has gathered at the mouth of a shingle stream. Somewhere in the woods a woodpecker hammers. The dawn is replaced by ordinary day; the emptiness soon filled by the commonplace. I can see my hands once more. Everything seems to pause in these final moments, as though the performance were put on hold, even as it begins again.

———

The beach isn't much of a beach. It's really all that's left behind by the slow-moving estuary, more a kind of watery cul-de-sac, fed by two converging rivers. One is filtered through chalk downland to the north-east, flowing through watercress and filled with swerving trout, slowly widening and losing its virginity until it reaches a carved-out bay in the semi-industrial arse-end of the city. Its outer curve, bulwarked by great heaps of rusting cars, is strewn with every imaginable item of litter, deposited by the tidal flow. Its fellow river emerges from the far side of the city, broadening through reed-seeded marshes into the shadow of the docks – a forbidden land where giant cranes stalk like wading birds, and where shiny new cars begin a journey which will end in the same kind of dump in the same kind of city. Yet somehow, somewhere, all this is forgotten in the conjunction of tide and shingle: something quietly miraculous, perpetually renewed.

The sea defines us, connects us, separates us. Most of us experience only its edges, our available wilderness on a crowded island – it's why we call our coastal towns 'resorts', despite their air of decay. And although it seems constant, it is never the same. One day the shore will be swept clean, the next covered by weed; the shingle itself rises and falls. Perpetually renewing and destroying, the sea proposes a beginning and an ending, an alternative to our landlocked state, an existence to which we are tethered when we might rather be set free.

I say it isn't much of a beach, but that doesn't do it justice, since it has a beauty all its own – more so for being seldom visited out of season except by dog-walkers and anglers. It is set at the end of a shallow bay that marks the south-eastern limits of the city. To reach it, I ride along a waterfront set with desultory concrete shelters and backed by common land to which are chained half a dozen horses. Behind stands a post-war housing estate, one-quarter of whose population live in poverty.

The path ahead passes through a stand of trees, then gives way to the beach, bordered by a waist-high sea wall with a narrow ledge, just wide enough for a person to walk along. To the landward is a sweep of grass and an avenue of oaks and pines. Gnarled and bowed, they mark an old carriageway that leads, with a grandeur out of all expectation, to a Tudor fort and a Cistercian abbey that once dominated this eastern bank of the estuary. Now the abbey lies in ruins, surrounded by scrubby woods and stagnant stewponds, while the fort, built out of stone robbed from the dissolved abbey, became a grand Victorian pile, recently extended in a replica of itself as a series of expensive apartments.

In this interzone, the modern world has yet to wipe out the past. Although the city is in sight, this place can seem haunted on a winter's afternoon, with its bare trees bent back by the prevailing sou'westerlies, and its rotting wooden piles, the remains of long-decayed jetties. The yacht club's boats stand unattended in their pound, the wind rattling nylon lines against aluminium masts in a continual tattoo.

Further down the shore, after an interruption of small shops and terraced houses, is a country park, the site of a huge military hospital. It exerted its own influence for a century or more, but it too has been demolished, absorbed into the turf, leaving only a nuclear shadow and its chapel dome poking through the trees. One day those tower blocks on the shore will be romantic ruins, too, relics of the work of giants.

If the weather is good, I'll cycle past the park to a far beach, overshadowed by holm oaks rooted in a low bank of gravelly, gorse- and bracken-clad cliffs, where southern England is slowly collapsing into the sea. As I ride, my route parallels the forest on the other side of the estuary. There may be barely three miles between me and its purple heath, but they might as well mark a continental divide. Not only because we are separated by water, but by virtue of what stands on the distant shore and now dominates the entire waterline, a triumvirate of new, industrial installations: oil refinery, chemical plant, and power station.

In the changing light, this cluster of cryptic structures could be anything. Tapering spires for a new place of worship; circular tanks as giant igloos, pale green with rusty streaks; silos like newly-landed space ships; tripod gantries ready to fire salvos of secret missiles. At dawn or dusk, the whole place might be a martial Manhattan, replicating every day, sprouting out of the shore, an alternative new forest of steel. There's no human scale to this petropolis; it has a curiously temporary feel, although it has been here for half a century. It might be disassembled in a day and imported to some other shore on the other side of the world. Stripped down, utilitarian, it makes no apologies to its surroundings. It has only one function:

to make the fuel that confirms its existence. It is brutal, practical, inevitable.

Like the nearby docks, this great complex, which employs more than two thousand workers, is sealed from public access. My own uncle worked there after his war-time service in Kenya, although I couldn't tell you what he did – any more than I know what he was doing in the middle of Africa in 1943, beyond the photographs of him in khaki shorts and a pith helmet, along with aerogrammes sent out by his family and kept carefully preserved in a Senior Service cigarette tin.

I've grown up with this place, which is only just older than me; I have no memory of the virgin shore before the coming of the towers, now imprinted on my view of the shore. Their stacks occasionally burst into life like huge Bunsen burners, as though the whole thing were some gigantic experiment, or a memorial to an unknown war-rior. First lit in 1951, these flares commemorate an age of energy and industry, power and destruction. Their function is to burn 'excess gases', but as their orange-red tongues lick the sky, they could be drawing directly from the molten depths of the earth, rather than the crude oil from the Middle East which is pumped from tankers that line the mile-long terminal, five abreast like petro-cows waiting to be milked, their bridges branded with slogans, NO SMOKING and PROTECT THE ENVIRONMENT.

To process its daily quota of three hundred thousand barrels of oil, Fawley sucks three hundred thousand tonnes of water from the sea, claiming to return it cleaner than it was before. (In fact, many living organisms are drawn in too: fish are caught in screens and often die, while smaller fry are sent through the factory's cycle as though through a washing machine, a process which few survive.)

Indeed, the word 'refinery' itself is deceptive, since its end products have precisely the opposite effect on the world into which they are released. And while the site is declared to be perfectly safe, its neighbours live in the knowledge that in the event of an emergency, they would be evacuated from their homes, just as the islanders of Tristan da Cunha were evacuated from theirs during the volcanic eruption of 1961, and were brought to a military camp here, in the shadow of the neighbouring power station, close to where their descendants still live. Its cylindrical chimney, an enormous ship's funnel on a concrete liner fading into pale blue, marks the end of the estuary and the beginning of the sea. Beyond is the tantalising outline of the Isle of Wight and its fields and downs. In the winter, when the trees lack leaves, I can see it from my window, its pairs of red points blinking like landing lights, foreshortening the distance between me and the sea, making it seem suddenly nearer.

Despite its industrial installations and historic layers, this is an unspectacular, unremarkable landscape. You could easily drive by and take away nothing from this place. Many do. No one writes books about this shore. No one who does not live here knows anything much about it, and even those who do would be at pains to report anything noteworthy about the place they see every day.

I just happen to live here. I didn't choose to; it chose me. I might have found a more picturesque place, wild and romantic or urban and exciting; the kind of places people pass through here to reach. A port city relies on its relationship to elsewhere. Perhaps that's why I like it so well, since it does not impose any identity on me. I came back here from habit as much as choice, like the birds that migrate to and from its nondescript shores.

You assume you know your home. It's only when you return that you realise how strange it is. I first saw this beach half a century ago, but all those years have made it seem less rather than more familiar. I've taken it for granted. But now, as I look out over its expanse, it occurs to me that what I thought I knew, I didn't really know at all.

The first recorded settlement here was Roman – the port of Clausentum – followed by Anglo-Saxon Hamwic; Southampton translates out of Old English as 'south home town'. Sholing, where I live, barely existed until modern times: its name is a contraction of 'Shore Land'; its neighbour, Netley, means 'wet wood'. Until the nineteenth century, this was common land, coursed by a Roman road and scattered with tumuli and cottages; an oddly isolated area, separated from the rest of Hampshire on three sides by rivers and the sea. Troops trained here; shanty towns of huts were set up for navigators working on the railway line and the sprawling military hospital. Their presence may have been why this place became known as Spike Island, a slur on its itinerant population of travellers and horse-traders, and a wry reference to a notorious penal colony of the same name in Ireland's Cork Harbour, Inis Pich. There was a wildness to this heathland. One corner was named Botany Bay, after the destination of the transportees who were held there – as their Irish counterparts were in Cork Harbour – before being shipped out to the ends of the earth.

Even now, this eastern side of Southampton Water can feel insular, outcast. A place through which to pass, rather than to stop at for its own sake. There's a sense that any-thing could happen here and already has, caught up in the flow of changing tides and people and animals and begin-nings and endings, the obscure currents of history.

Recently, I flew home after spending some days on the banks of a Scottish sea-loch. The north had been dramatic, monumental, with rivers of mist running down gran-ite valleys, falling like dry ice into the still, deep water from which the mountains rose on the other side, blue-purple and faintly oppressive. The skies were overcast by the damp Gulf Stream and second-hand winds from the Caribbean. Flying back south, I felt an immense lifting of the atmosphere as the sun broke through the clouds and the plane banked over Southampton.

In those few minutes I saw the past and present unroll beneath me, as in a camera obscura. The horizon had vanished, to be replaced by a careering view, as chaotic as it was ordered. The plexiglass porthole filtered the light like a prism, and gave the scene a watery air. I might have been looking out from the side of a ship or even from a submarine.

Down there, somewhere, was my house, sheltered by trees and shrubs. To the south lay the sea, a broad strip of blue-green bordered by yellow shingle. As the plane flew over the beach I knew so well, but could hardly recognise from this angle, it turned back from the forest and into the heart of the docks, passing monumental cranes and ocean liners lined up like bath toys, swimming pools shimmering in the sun. Then it flew over the bridge that connects one side of the city to the other, over the school I first attended more than four decades ago.

The afternoon sun lent it all a glowing sheen, highlighting every reflective surface. Even the scrapyard and its defunct cars reduced to metal screes acquired an allure, shining like piles of iron filings. Seeing all this laid before me, even after only a few days away, made me intensely happy and profoundly sad, its streets and shores so familiar that they seemed extensions of my own body. Finally, descending to the airport, we touched down on southern soil once more.

This suburban sea is a living thing, ever shifting as it is contained. Everything seems open to the light, some subtle combination that has never been seen before and will never be seen again: the sun forced from under a bank of cloud, a pure white egret like a flapping apparition, a pair of mute swans gliding close to shore. Even on the dreariest days, the most forlorn afternoons, it's never not beautiful here. The slow surging waves seem to be suppressed by the mist, yet every sense is heightened. I can smell the forest across the water. Sound behaves differently; with no buildings to bounce from, it spreads over the surface and soaks back into the sea.

Black-headed gulls, barely more than pale smears, splash their heads and wings. Unresolved shapes drift by. Everything coalesces, caught in a dreamy, half-hallucinatory loop. There are shadows under the water as it withdraws over weed and rusty outfall pipes. A distant yacht becomes a silent white smudge. A fall of black crows scatter in the murk. The saturated greens and browns of grass and leaves turn the colour balance awry, in the way that reds and greens take on an eerie vividness before a storm, when the pressure appears to affect the light itself.

My time is determined by the ebb and flow. At low tide, the beach is an indecent expanse laid bare by retreat, more like farmland than anything of the sea: an inundated field, almost peaty with sediment, as much charcoal as it is sludge.

Bait-diggers leave little piles by their sides like slumped sandcastles. With their buckets and spades, they might as well be burying as disinterring, these sextons of the shore. Standing over them is an outfall marker in the shape of an X, which has turned to become a cross. In the uncertain light, the mud takes on new colours, from black to taupe and even a kind of rubbed silver.

As I say, it is never not beautiful here.

Behind me, bare oaks and beeches lie as cracks against the sky, evoking a peculiarly English landscape. In the late eighteenth century, Turner drew the abbey's ruins in his sketchbook, tracing out the trees that had grown up around the crumbling gothic arches. In 1816 John Constable stayed at Netley on his honeymoon, and painted its scudding clouds and billowing greenery.

Theirs were records of a Romantic setting, an alternative reality of sensation and emotion. Hanging over the shore, the gnarled, enamelled branches are made darker by the reflected light of the sea and the stretch of bright shingle. My shortening eyesight renders it all abstract, blurring the scene like Turner's evocations of nothingness, with their vague shapes that might be waves or whales or slaves tossed overboard, rising and falling in foam with 'a sort of indefinite, half-attained, unimaginable sublimity about it that fairly froze you to it', as Ishmael says in *Moby-Dick*, 'until you involuntarily took an oath with yourself to find out what that marvellous painting meant'.

That fixity of sea and sky is a supreme deception. Over it lies what Herman Melville called the ocean's skin – a permeable membrane, one-sixteenth of a millimetre thick, fertile with particles and micro-organisms and contaminants; a fantastically fragile yet vast division. The horizon is only an invention of our eyes and brains as we seek to make sense of that immensity and locate our selves within it. The sea solicits such illusions. It takes its colour from the clouds, becomes a sky fallen to the earth; it only suggests what it might or might not contain. Little wonder that people once thought the sun sank into the sea, just as the moon rose out of it.

Not many artists come here now to see the sun set or the moon rise. Netley's beach is hardly thronged with easels, and Turner and Constable left long before the refinery turned the shore spiky with petro-chemical romance. Perhaps the strangest thing about this massed industry is its absence of sound, at least at this distance, both innocent and ominous at the same time, although occasionally the plant emits a dull indefinable roar, like a giant stirring in its sleep.

This is a place both dead and alive. Being here, in or by the water, at either end of such a cold, closed-down day makes me physically part of it. A crested grebe pops up, charting the shallows for its prey. It arches its neck to dive again, as I swim towards it. A jumpy, almost nervous plunge, a little leap forward, then it's gone.

As with so many things about birds, you have to take a lot on trust; to discern what is real, and what they want you to believe. The pattern of a duck's feathers, for instance, breaks up the surface, blurring its body between water and sky, an effect that prompted the American artist Abbott Thayer to suggest eye-dazzling camouflage for First World War soldiers and ships, mimicking natural cover in an industrial war. A grebe's markings, intensely detailed against the grey of the sea, may seem darkly obvious to us, but from below, its white throat and breast renders it invisible, able to deal death with its stiletto bill. Thayer called this disguise 'counter-shading'; it lends animals from birds to whales a flat deception in the overhead sun, making them seem insubstantial rather than solid things.

As they emerge from their dives, I realise that there are half a dozen birds patrolling, moving into their winter haunts. From their own level, in the water, I feel accepted, or at least tolerated – any bird will have seen you long before you see it. I'm too far away to see the grebe's blood-red eyes, although there's a suggestion, even at this distance, of the outrageous ruffs which once supplied Edwardian society with stolen plumage. But then, grebe parents will pluck out their own feathers to feed their chicks, lining their young stomachs against fish bones.

The tide ebbs, and the birds assemble, as if someone had laid the table and called them in to dinner. Gulls and geese are already working the shoreline, as are the oystercatchers, in their white winter 'scarves'. They're one of my favourite birds: familiar, stalwart, forever looking out to sea.

Untrue to its name, imported from its American cousin, the Eurasian oystercatcher eats mostly mussels and cockles teased from the shore, using its greatest asset: a bone-strengthened bill, part hammer, part chisel, able to prise open the biggest bivalves. Delicately coloured carrot-red to toucan-yellow – it might be made of porcelain – it is a surprisingly sensitive probe. At its tip are specialised Herbst corpuscles that allow the bird to sense its prey by touch as well as sight; an oystercatcher can forage as well by night as by day. Perpetually prospecting the beach, it stabs and pecks or 'sows' and 'ploughs', altering its methods to suit its prey. It can even change the shape of its bill – the fastest-growing of any bird – morphing from blunt mussel-blade to fine worm-teaser in a matter of days.

Living and dying by its wits, the oystercatcher has evolved to take advantage of its environment. It has been present on Southampton Water for centuries, if not millennia: graffiti'd into the sixteenth-century plaster wall of the port's oldest house is an oystercatcher, a scratchy Tudor cartoon of an animal familiar from the nearby shore. In 1758, Linnaeus classified it as *Haematopus ostralegus* – blood-footed oyster-picker. In Britain, it was prized as a dish (although its name, 'sea pie', actually came from its piebald colour) and reduced to near-extinction in some southern sites. In modern times the birds were seen as threats to cockle beds: from 1956 to 1969, some sixteen thousand were shot in Morecambe Bay alone.

My oystercatchers, if you'll forgive the proprietary tone, may feed and roost here, but they nest as far away as Belgium and Norway. They'd do well to steer clear of France, where hunters still shoot two thousand of them every year for recreation, sometimes ten times that many. These monogamous animals have complex social structures and can reach forty years or more, the longest-living wading bird. And they always return here, where they feel most at home. Their peeping calls drift over the water as they fly in low formation, wings emblazoned with white streaks. They settle to forage on the tideline, occasionally breaking into indignant arguments.

I raise my binoculars. I spy on them, they spy on me, one eye always on the stranger. As I watch, they're joined by ringed plovers. The skittish newcomers bank in synchronised circles, suddenly swerving as if they'd hit some unseen current, then performing a deft communal turn to land. As they do so a gull takes off clumsily, lurching into their flight path and causing them to scatter. Rapid wings rev into reverse. All of a sudden, they're on the ground, camouflaged bodies merging into the mud.

As timid as they are, the plovers are unfazed by the ever-present carrion crows that have established themselves here, a black-flapping backdrop in the car park. Overnight they'll empty the bins like delinquent dustmen,

leaving the tarmac strewn with the guilty evidence of their scavenging. Surveying the drifting fish-and-chip wrappers, they avert their eyes as if to say, 'It wasn't us.' Although my books tell me that they're solitary animals, they gather here in a great fluid flock of two hundred or more. Perhaps they're evolving into aquatic birds, just as the gulls have moved inland to rubbish tips and shopping precincts.

There are other worlds of communication going on here, unknown orchestrations of action. Every so often the crows will rise up in waves, bird-shaped holes in the sky. They're a lustrous lack of colour, denuded of detail; a fluttering negation, as dark as the night. Ted Hughes, who made a new myth of the crow, saw the bird as suffering everything even as it suffers nothing. Encouraged by its ugly name, we indict its assemblies as 'murders'; yet crows mark the passing of one of their number in funereal demonstrations, cawing their grief in the way elephants and whales mourn their dead.

These ignored birds – whose ubiquity only makes them less visible – display the fascinating behaviour of their family. Broad-shouldered males swagger from one leg to another. Using their thick, oiled-ebony beaks, they peck over stones so much more dexterously than a wader or a gull. There's a determined, discretionary air to their

epicene foraging, although actually they'll eat almost anything. They seem surprised if you stop and look at them, as though no one had bothered to do so before. They stare back briefly, abashed, then turn away, unable to believe that anyone other than their own might find them interesting. Or perhaps there is disdainful pride in that sideways glance, assuming the reverse: that they are the most intelligent of all birds.

As indeed they are. Raptors may be more majestic, songbirds sing more sweetly, waders are more elegantly poised, but corvids such as crows and ravens exceed them all in matters of the mind.

You can see it in their body language. They're full of character, with their grizzled, quizzical stances; individuals, possessed of particular attributes. Their eyes glitter and their heads swivel with curiosity, ever alert to what is going on around them. Bold and twitchy, timid and territorial, their restlessness is a sure sign that something is going on in their heads. Singly or en masse, they react to every sound and movement. They're always aware of what the others are doing: fighting, preening, competing, conspiring, minding each other's business to see if it can be outdone. If a fight breaks out between two of them, the others will swoop in from the trees around to see what's going on, like children in the playground chanting *Fight, fight, fight.* They're irredeemably nosy, socially-adjusted birds.

Crows appear crafty to our eyes, since we seem to find intelligence in any other species than our own suspicious (I write all this down in my policeman's notebook, as if I were about to arrest one of the avian young offenders).

They're an alternative community over our heads; gypsy birds, a mysterious race with their own hinterland. They live on the periphery in the way that all animals do, existing on the same plane as we do but inhabiting another time and space. They even have their own voices, resembling the patterns of human speech: captive corvids can be taught to speak as well as, if not better than parrots; it is one reason why they were said to be carriers of dead men's souls. Acting in loose unison, at some unspoken signal they will fly out of the woods and onto the shore, as if they were the spirits of the monks evicted from their dissolved abbey. No one really knows what they do or how they think. Perhaps theirs is just a convenient congregation, only motivated by food and sex. But then, you might say the same about us. As a species, we are unable to resist the temptation to impose our own failings on animals; it's almost an act of transference, and I'm as guilty as anyone else.

When the water has fallen back far enough, the crows will swoop on shellfish, rising up to drop them on a stone from a perfectly judged height. All the while they keep one eye on their fellow birds, ever ready to steal from friends or passing squirrels. They're a disputatious, bullying, larcenous lot, forever finding fault with one another. They'll tumble two-against-one in aerial combat, before falling to the mud to scrap over a mussel, the soon-to-be-loser on its back, eyes glaring, claws defensive, determined not to let go of its hard-won bivalve. Then, as suddenly as it started, it's all over. A moment later and the same birds are strutting alongside each other perfectly amicably.

They may be vermin to most people, but I've come to love carrion crows. Sleek and knowing and iridescent, they could be in disguise for all I know, glossy agents sent to spy on us. If I were to die here on the beach, it would be the crows who picked my bones clean.

Even on this nondescript stretch of water the colonisation continues: the annual invasion arriving over here, and the annual exodus leaving for over there. Around now the dark-bellied brent geese appear from Siberia; one-tenth of the world population winters along this coast. For them, Southampton Water is one big runway. Travelling three thousand miles in six weeks, they're long-haul flyers, built to purpose with compact bodies, sturdy ringed necks and neat dark heads. I hear their rolling honk as they pick their moment to settle, their voices eliding and trilling like the chorus from a minimalist opera. Even their name sounds northern: 'brant' is Norse for burnt. As the tide flows, they will ride the surf like little ships, proud of their survival, joining the herring and black-headed gulls hunkering to the swell; at this time of year, the air is colder than the sea.

This international, modest gathering of birds – constant and ever-changing, unremarkable and exquisite – are united only in their search for sustenance. I have to remind myself that they're not here for my entertainment. They choose this part of the shore because it is a fertile patch, fed by a freshwater stream that oozes from the woods, turning brackish in a holding pond before running clear to the sea, and with it the nutrients that feed whatever the birds feed on.

They're always waiting for the tide to go out; I'm always waiting for it to come in. Time may move faster for them – a day to them is a month to me; they live ten lives to mine – but they've been here for generations. This is their refuge. They feel safe here, despite the bait-diggers who disturb their foraging and the heavy metals and organo-chlorines that pollute their food and threaten their fertility. They remain loyal to these blackened flats; northern animals, like me; philopatric, home lovers. They were here before Jane Austen came to visit the abbey's ruins, and before the Romans rowed up the river to establish their own colony.

We should pay attention to birds, says Caspar Henderson; 'being mindful of them, is being mindful of life itself'. They have always surrounded us; our movements mirror theirs. For humans, this too is a place of migration and emigration: from the tribes who came here when the sea was still a river, to the post-war 'ten-pound Poms' sailing for Australia – among them my school friends, never to be seen again – to the Filipinos, Poles and Bangladeshis who constitute the city's latest arrivals.

What if all the vessels that ever sailed this water rose up from this much-dredged ditch? Celtic coracle, Roman galley, Norse longship, Tudor barge, Victorian merchantman, interwar liner, twenty-first-century ferry, all tumbled together like Paul Nash's painting of dumped war planes, *Totes Meer*, Dead Sea. Sometimes the trawlers spit out fragments. I've found rusting revolvers brought back by wartime troops and chucked overboard when they were forbidden to import their souvenirs. Two thousand years before, their Celtic counterparts cast offerings to the water gods, and Roman centurions tossed tokens to Ancasta, the river deity who lived in this estuary. It all lies there, entire worlds of marine archaeology awaiting excavation; crockery and weapons and bones piled in a watery midden.

Out of the calm there's a sudden surge as if some invisible vessel had passed by, followed by a reluctant riffling, running over stones as waves first set in motion thousands of miles away spend themselves on the shore. The sea plays its own tricks here. For two hours or more, its height and weight is suspended in a delayed action produced by the Atlantic pulse, although I must admit that the logistics of this mechanism somewhat mystify me.

If I understand it correctly, the tide runs from west to east and back again, courtesy of the pull of the moon, rocking up and down the Channel like a seesaw. Set at its midpoint, Southampton's tide bounces back up its estuary. But added to this are local complications. The stopper of the Isle of Wight creates further oscillations, as the water enters and leaves from either side. The result is a unique selling point for the port. For centuries this double tide has been a boon to marine traffic, making Southampton 'a seaport without the sea's terrors, an ocean approach within the threshold of the land', according to one nineteenth-century account. Its downside is what it leaves behind, an intractable stretch of mud, scattered with debris.

This is a place with its own rules. Its performers enter and leave the stage left and right, from wading birds pecking at the mud to slow-moving tankers pulling into the refinery to be suckled dry of their tarry cargo, and flat barges bearing turbine blades on their backs like sleek grey cetaceans. But they're all dwarfed by the estuary's most evident yet oddly ignored actors: container ships and car carriers.

Registered in Kobe, Panama or Monrovia, their names aspire to a Western status – *Lake Michigan*, *Austria*, *Heritage Leader* – while their sides are proprietarily stamped Maersk, Hapag-Lloyd, Wallenius Wilhelmsen, or with the anonymous initials – NYK, EUCC, CMA–CGA – of commercial states. Apart from the fact that they float, there's little to associate these giants with the romantic notion of a vessel. Rather than roaming the seas, they're locked into rigid routes. They accomplish in days journeys that James Cook took years to traverse. They're standardised to the width of the Panama Canal: ships made to fit a world made to fit them. They might as well have been chopped off the production line. Their cantilevered prows look down on everything else, but their square sterns appear wrinkled, as if they were papered-over hardboard.

No one rhapsodises over these maritime pantechnicons as they come and go on their migrations. No one celebrates their arrival after heroic journeys to and from the other side of the world. They are filled. They are emptied. They move in between. No one stands on the quayside to wave them off. There are no Royal Marine bands to see them on their way. No bunting, no ceremony, no joy or sadness, just a slipping away. They embody a shrinking world. Half as big again as *Titanic*, they sail down the same waterway with bland indifference, lateral tower blocks so huge that, as one waterside inhabitant tells me, they cut off all electronic signals as they pass by. Their sides dribble with rust; the sea will get them in the end. They are ghost ships, devoid of life, save for shadowy figures seen through letterbox slots let into their flanks.

A life at sea? Their ill-paid crews might as well have signed up to a sweatshop. Since the decks are too dangerous to walk, the men remain within the metal hulks, themselves contained. Yet these ships carry almost everything we consume. Top-heavy, stacked with blocks like toy bricks, fifteen thousand tonnes of steel, four hundred metres long, sixty metres above the surface and another ten below, they ride high in times of unequal exchange. As one captain says, 'We take air to the East.' In better circumstances they sit lower in the sea, a plimsoll line of the global economy. At the end of their journey they will be unloaded by bestriding cranes onto railway trucks. In turn, others take on shiny new vehicles shelved in multi-storey stacks like factory-farmed chickens, from which they are driven into the bellies of the floating car parks and out the other side to Singapore.

But then, this is a city of the sea, built on reclaimed land; even its railway station platform is composed of cement and shingle embedded with shells. Meanwhile those same ships bring back invasive species on befouled hulls or in their ballast, Japanese seaweeds and Manila clams; to marine biologists, this is probably the most 'alien' estuary in Britain, with new organisms arriving every year.

In the nearby National Oceanographic Centre – an oddly industrial-looking complex itself, mounted with telecommunication masts, and flanked by long refrigerated sheds that contain sample cores of the earth's archival depths, every three metres representing one hundred million years – I study the Admiralty Navigation Charts of these commercial waters. Pulling great plasticised sheets from chest-high cabinets, I pore over maps that have turned the world around to show the importance of the sea to the land, rather than the other way about. Atlases display the waters around our coast as blank blue expanses, but here all the contours and depths are laid out, along with their utility to men and women at sea.

The shore from which I swim, for instance, is labelled, unappealingly, 'East Mud'. Nearby is a 'Swinging Ground', along with a 'Hovercraft Testing Area' and 'M.O.D. Moorings'. Buildings, houses and roads have vanished, to be replaced by sites selected only for their relevance to the sea. Church spires and 'Tall Buildings' become landmarks, identified by bald details: 'House (red roof)'. Through Southampton Water and into the Solent and the Channel beyond, a martial arena is mapped out: from the benign 'Dolphin Bank' to the treacherous shallows of 'The Shingles'; from 'Radar Scanning' and 'Foul Area' to 'Firing Practice Area', 'Submarine Exercise Area', 'Explosives Dumping Area', and 'US Base'; the reverberations of seismic global conflict brought to placid inshore waters. They are designated training grounds for the closed installations scattered along this coast, like the military port at Marchwood, busy sending ships laden with materiel to foreign wars and bringing back the broken remains. After the Falklands war, the bodies of eighty men were stored in its cargo shed.

29

One afternoon, after sitting on the sea wall watching the birds, I was about to ride home when I saw a strange shape moving down the water. It sat low on the surface, matt black, absorbing the light around it. Escorted by three tugs, the nuclear submarine – HMS *Tireless*, a 'hunter-killer' here on a 'friendly visit' – glided slowly south, powered by invisible force. I could see figures on its conning tower, and others walking the length of the vessel. They looked precarious to me, moving down its rounded back with no restraining railings to stop them rolling off; they might as well have been strolling on the back of a whale. As I watched, two of the crew reached its high tail fin and from the vessel's stern pulled out a white flagpole that stood there, as though it were a parade ground. It was being prepared for its descent.

Soon, somewhere off the Isle of Wight, it would submerge into the English Channel, and travel six thousand miles beneath the surface of the Atlantic to the Falklands. Nuclear submarines are so efficient that they can stay below for three years or more. In Scotland, a taxi driver told us how he'd worked in Faslane, at the submarine base. He said that the submariners' mail was habitually screened for any possible bad news from their families which might cause them upset. Even if their loved ones had died, there would be nothing they could do about it – there'd be no return to shore.

The driver spoke in a matter-of-fact manner of men going mad at sea, losing their sanity in the confines of a metal tube where they might not even have their own bunks, but be forced to share beds in sequence with their mates. He said one man had appeared in his civilian clothes, carrying a bag, saying he was ready to go home now.

One morning I arrive at the beach to an extraordinary sight, so unexpected it causes me to screech on my brakes. The water has disappeared, to be replaced by mud flats. It's as though the plug has been pulled on the estuary, and an entirely new landscape has appeared. In the extreme spring tide, the channel has been reduced to its absolute minimum, so narrow you might almost stroll across to the forest.

Posts rise out of the mud like dead men's fingers, ready to pull me down as I try, unsuccessfully, to walk out to this new world. The birds have it all to themselves. Even the crows have turned their backs on the human world in which they scavenge and are off in the distance, bathing with the waders.

The tide itself is weather. The weak sun tries to burn off the mist, but it only gets colder. There are astonishing effects in the sky, reflecting the sealessness below. It's like being in an eclipse. Perhaps the river Solent is about to return to its antediluvian state, or perhaps this is the precursor of a freak tsunami. Or maybe the sea has relocated to the sky, as it was once thought there was another ocean over our heads. One medieval chronicler related how a congregation came out of church to find an anchor snagged on a gravestone. Its line ran taut to the clouds, from which a man descended, only to be suffocated by the dense air as if he were drowning.

Huge yellow buoys which normally float from chains that anchor them to the sea bed lie slumped like giant beach balls, left behind after a day's play. At the dockhead, ships' flanks are indecently exposed, as though someone were looking up their skirts; unsupported by the water, they might fall over at any moment. But the withdrawal must stop at some point. Soon normality will resume, and the earth and the moon will go on turning, tugging the sea between them. Some days, in late autumn, the fog is so thick that the sea and sky merge into one. There may be hundreds of birds around me, but I only hear their squawks and peeps. Unseen ships moan like lost whales.

Winter closes in, sweeping the mist away with Arctic winds. The air is so cold it seems to crack the tarmac. My fingers turn raw and crab-like; the colour of summer has long since faded, leaving brown islands on the back of my hands. It's time to start wearing two hats, as well as two pairs of gloves. Shoulders hunched, I push my bike along the beach, knowing full well that the water will be even colder. At this nadir of the year, people ponder the wisdom, or not, of getting out of their cars. For me it's all a question of getting in.

I stand over the water, and wonder why anyone would want to enter it. The surface is pressed flat by the cold. Slow and viscous, it wrinkles like setting jam. An oily sheen spreads over it. Rafts of usually active herring gulls float as if frozen into place. Everything has slowed to a gla-cial pace. Later the sea will ice up at the tideline, like the salt around the rim of a good margarita. In the summer, the water expands with the warmth; now it physically shrinks with the cold. Checking the coast is clear, I pull off my boots and my clothes and wade in without thinking.

I push through the waves with ice-cold hands. From above, I must look like a clockwork frog. My animal heat retreats with each forward stroke; I reach out as if to warm up the water. In summer, my body settles in comfortably; now everything is taut, demanding the conservation of its core.

I line up to the distant markers where cormorants perch. I've reached my limit. I turn back to the beach, scrabbling like a goose to find my depth once more. Naked on the sea wall, I give a little dance, singing to myself. If 'ecstasy' means to stand outside yourself, then I feel happier than I have ever been. Everything stripped away; everything renewed. Just me and the sea.

In the wan light the sun is diluted and dumbed. I struggle to put on my clothes, shivering as if the whole world were shaking, rather than me. My feet leave suspended puddles on the concrete, each toeprint in three dimensions. There are red threads from my towel caught in the cracks from earlier visits. I tug on my socks. Back home, I'll shake out the sand and weed as evidence of my folly.

The cold becomes a kind of warmth. My fingers burn as the feeling returns, like they did when I was a boy, home from school and holding my hands too near the gas fire; by winter's end my knuckles will be cracked and bleeding. With my heightened senses, I smell the lanolin in my woolly gloves. When I manage to scrawl in my notebook, its pages held down with an elastic band, my nose drips onto the ink, turning it into Rorschach blobs. My body complains of the lack of sleep. The prospect of tea and toast and a warm house never seemed so alluring.

Yet with all this self-imposed torture comes an intense, capillary clarity. Perhaps it's just the blood pumping back to my brain, but I feel as if something had been wiped clean. I'm ready to start again. I feel in the world, not just of it, even though sometimes, in the mist, I think I must be still dreaming.

Winter is a lonely season. That's why I like it. It's easier to be alone; there's no one there to notice. In the silence that ascends and descends at either end of the abbreviated day, there's room to feel alive. The absence makes space for something else. I must keep faith with the sea. Swimming before dawn, it is so dark that I have to leave my bike light on so I can see where I left my clothes. Once the waves washed them clean away, leaving me to wade after them.

The sea doesn't care, it can take or give. Ports are places of grief. Sailors declined to learn to swim, since to be lost overboard – even within sight of the shore – and to fight the waves would only extend the agony. You can only ever be alone out there.

People have died here, in these suburban waters. In the cemetery of Netley's military hospital, planted as an arboretum to blunt the edges of death, there's a gravestone carved in solid Cyrillic characters, a memorial to three Russian sailors from the frigate *Prince Pojarsky*, who drowned here in 1873. In the nearby pub, an outbuilding once stood as a temporary morgue for bodies pulled from the water by the coastguard, their corpses laid out on tables while next door people drank their pints of beer. My elder brother, working on a trawler off the Isle of Wight, once watched as the net pulled up a body, one of two men who'd decided to strip off at midnight and go for a swim. The fishermen kept the bloated corpse netted off their bow until the police arrived; it is bad luck to have a body aboard a boat. Like those unswimming sailors, I can't reconcile my love with my terror. I know full well what lies beneath me as I push out from the wall and into the water; and yet I still fear what it might contain.

One day, with the sea swollen by a near-full moon, I get the feeling I'm not alone. I've just turned back from my farthest point when I'm startled by a sudden *whoosh*. Directly behind me, barely a yard away, is a huge head with shiny dog-like eyes: a large grey seal, fat and full-grown.

I back off, shocked at the sight. I knew there was a seal colony just along the Solent – I'd seen grey and harbour seals there, lounging on the mud flats, so blubbery and lazy that algae grew on their backs where they spent all their time basking in the sun, raising their hind flippers in the air to keep them warm on chillier days. From a distance, they look quite cute. But coming face to face with one in the water was another matter. Weighing up to eight hundred pounds, grey seals have sharp claws and teeth that can cause a serious infection, *Mycobacterium marinum*, otherwise known as seal finger, which may result in the loss of affected digits.

The seal and I regard each other, equally surprised. He's twice my size, clearly a mature male. He raises his grizzly head, lugubriously. I'm not sure what he intends to do, but I'm not going to wait to find out. Kicking out with my feet to persuade the animal to keep its distance, I make for the shore – only to discover that the great beast has followed me, swimming beneath the surface. Scrambling onto the safety of the sea wall and reaching for my clothes, I look down at it.

I was right to be apprehensive. Up close, it is even bigger, almost magnified by the clear water. It looks more like a manatee as it hangs there, puffing away quite quizzically, all whiskers and wrinkles, trying to work out what I am, this pale, unsealish creature. I hurry to dress, keeping one eye on my marine companion. His curiosity satisfied, he turns towards the open water and sets off, popping up at intervals as he works his way upstream, before finally moving out of sight.

Back home, I walk around the house in the dark. I know its rooms as well as I know my own body. I catch myself in the mirror on the landing, hung so that my mother could check her make-up before coming downstairs, her necklace in place, just as my father always wore a tie. Now I look in it and wonder who I am.

I step outside, under the frost-sharpened sky, and a watery array: Pisces, Aquarius, Capricornus, Delphinus, and Cetus the whale; a starry bestiary (as if infinity wasn't frightening enough already) of ancient patterns created by minds yet to be overwhelmed by the images that fill our waking day. They fall in slow motion – Orion's brilliant grid, Betelgeuse's dying watch-jewel, the Pleiades' nebulous cloud – seen in the astronomer's averted vision, as if too big to look at directly. They seem unchanging, but they represent cataclysmic explosions, speeding into oblivion, collapsing into themselves.

The nearness of the sea opens up the sky. I hold my binoculars shakily to a three-quarter moon, its cold face forever turned away; to Sylvia Plath, it seemed to drag the sea after it 'like a dark crime'. Once, out in the garden late at night, I watched an unusually bright meteor flashing orange, red and white. As it fell to the horizon, its tail streaming behind like a medieval illumination, I heard it hiss and fizz.

Far off in the city centre a clock tower chimes. Inside the house, things shift and fall. Floorboards creak like a ship. It ticks with the ebbing heat as it falls asleep. I lie in my narrow bed, listening to the sound of the dark. A vague rumbling drifts over from the docks, godless, twenty-four-hour places where the black water ripples with sodium traces. Turning off my bedside light, I hear someone call my name, as if the night won't leave me alone. Evenings I once spent drinking and dancing and taking drugs are now filled with a heady emptiness. Late at night, I think there's some animal stirring in one of the rooms, a bear cub being licked into shape. And sometimes I wake in the early hours to hear my mother washing up downstairs, even though she died six years ago.

The house has its own history, plastered over, extended, reduced, rising and falling with fashion like the hemlines of a woman's skirt. The lawn where I lay as a teenager, reading *King Lear* on a hot midsummer's afternoon although I'd rather have been listening to *Ziggy Stardust* on my cassette recorder, has long been overtaken by meadow grass. Somewhere deep in the bushes is the chain-link fence that first marked these plots parcelled out on the heath by a 1920s developer. If you can age a hedgerow by the number of species in a given stretch, you can date a street by its styles and details. Things here were once more empty: open coal fires rather than central heating, a hot-water geyser that exploded into blue life over the old enamel bath, and a bare electric fire hung even more dangerously overhead. No telephone, no fitted carpet, no double glazing; children spilled out of doors.

Then the view was open to what lay ahead, and a shop stood on each corner of the crossroads: a grocer's and post office combined, where you could buy postal orders while your luncheon meat was sliced; a butcher's shop with tiles and sawdust and bloody lumps of offal; a hairdresser's with its oval helmets that made their occupants look like astronauts, preserved in permanent lotion; and a brown-painted cubby-hole of a shop run by a lone elderly lady which sold only sweets and was rarely if ever open. All gone now. Here, as elsewhere, suburbia has disappointed its utopian dreams. Bramble finds its way into every crack.

At the bottom of the garden – beyond the summer house whose interior is festooned with ancient spiders' webs, each dangling white drop holding a mummified fly – is a crumbling potting shed. Recently, a sudden hailstorm caught me out in the garden, and I dived into the shed for shelter.

It was the first time I'd set foot inside the place for months, maybe even years. The roof was rotten and yawning to the skies in two places, as though a bomb had hit it; everything was decaying with lost summers and long-dead flowers. A pair of deflated bikes stood stacked against the tilting walls. Plant pots tottered in towers. Bamboo canes which once guided sweet peas to the sun gathered in the corner, the twine still wound around their knotty rings. The entire edifice was slowly decomposing. As I stood there, still in the silence save for the rattle on what was left of the roof, the hailstones poured in through the holes like sand in an egg timer, threatening to fill the interior with granulated ice.

Behind the shed a high privet hedge, blowsy like green clouds, hides an alleyway where my brothers would collect grass snakes, slithering in a bucket. Hedgehogs still shuffle out of it at night, leaving paths in the grass. At the end of the cutway, across the road, lies what is left of the common, a narrow, tree-filled valley dipping down to a stream, from where I hear the call of a tawny owl at night, drifting over the roofs, as if it might be caught by a satellite dish.

One afternoon my father, working on his much-manured vegetable patch, called to us to see a slow worm on the lawn; it must have slithered out from the compost heap. For some reason, I picked up a spade and drove it down on the lizard, slicing it in half. I remember being surprised by the blood that oozed out of the bisected reptile.

What had I expected? Did I think it was one of my rubber toys? Fascinated and horrified by what I'd done, I stood there, staring stupidly. I still regret it. On only two other occasions have I been personally responsible for the death of an animal. One was a hedgehog that I once found with a growth on its eye like a bloated pale pea. I drowned it in a bucket of water, holding it down, feeling its tight-balled body open and close, briefly. The last was rather more recent.

I first saw it out of the corner of my eye, a white blur on the beach. After I'd swum I saw that the bird was still there.

It was nothing unusual: a black-headed gull. As I walked towards it, it ran off rather than took to the air. Then I saw its wing tip hanging, obviously and dramatically broken. Had a dog or a fox done that? Every now and then it tucked its head round to the injured site, pecking at the disconnected bones, unable to understand. Why couldn't it do what it usually did? It could not comprehend the malfunction. I could, but I ignored it, and cycled home, thinking determinedly about lunch instead.

A year or so before, I'd been cycling along the beach when I saw a pair of mute swans, a common enough sight here, running their belisha-beacon beaks through the shallows. The Nordic historian Olaus Magnus wrote in 1555: 'The swan, as everyone certainly knows, is a placid, good-natured bird.' He noted that it derives its name, *Cygnus*, from its singing, sounding sweetly from its long, curving neck, although he added that in old age, it sings with one wing over its head, its swansong 'as it departs from life. Plato says that it sings not from sadness but from joy as its end draws near.' Like the ermine, said to prefer to die rather than soil its white winter coat, the swan is pledged to maintain its whiteness, as immaculate as the newly laundered shirt of an African schoolchild.

But one of these swans was not preening itself. It was tugging with its wings at a near-invisible thread: a micro-filament of discarded fishing line which threatened to trap the bird till it was trussed up in its own panic.

A man with a weather-beaten face was also looking on, concerned. I suggested we try and do something. We waded out towards the pair, but they moved off, out of reach. My comrade, as I now regarded him, made a suggestion. He owned a rib, and would fetch it from the pound. Moments later we were in the boat, in pursuit of a swan. The powerful outboard motor took us out into the channel. Our prey, meanwhile, was doing its best to evade us. Circling to cut off its escape route, we drew close to the entangled bird. Ignoring what I'd been told – how a swan can break a human limb with its wings – I leaned over and gathered it up in my arms.

I felt its whiteness in my embrace, sturdy and warm and downy. It didn't struggle at all. Indeed, it seemed quite at home, although that may have been merely fear: captured animals will play dead, as a last resort. It was like holding a living musical instrument – or, perhaps, putting one's arm around a ballerina. Its head swivelled to face me indignantly. I thought of Alice's flamingo mallet as I gripped the slender, muscular neck. My fellow res-cuer pulled out his penknife and cut the line. It was all over. I opened my arms and felt the bird's tension release and the life return to its body. It swam off for a few yards, stretched its wings, turned, and honked.

Only later, reading Tag Barnes's *Waterside Companions*, would I see another side of the story. Barnes, an angler since he was a boy, wrote his book in 1963 to enlighten fellow fishermen about the creatures they might see around them. Moorhens, cormorants and grebes all get their admiring page or two; even water voles, toads and coypu are given their due. But when it comes to the swan, Barnes appears to lose his temper: 'I simply bristle with annoyance when one moves into my swim.' He certainly does not agree with Plato's lyrical tales. Swans, he says, are 'the most aggressive, persistent and arrogant birds we have and can be extremely dangerous'.

It was refreshing to read natural history from the predator's point of view; in Barnes's words the mute animal becomes almost malevolent. No amount of stone-throwing or shooing would deter these waterborne thugs, he says. 'They will often hiss back at the "shooer" and sometimes threaten him with violence.' He suggests that dirty water might be thrown over the miscreants – perhaps on the basis that these vain creatures would be humbled by their besmirched plumage. His other remedy is to 'cast a line over their backs'. Is that what happened to 'my' bird? Barnes convicts himself in his final blast, as deliberate as the second discharge from a twelve-bore. Having admitted that swans eat only aquatic plants, he concludes: 'I can recommend the cygnet as being a really tasty dish!'

Faced with the plight of the gull and its broken wing, I returned to the beach that afternoon. I'd last seen the bird running into the bushes, seeking shelter from preda-tors; I'd watched other injured birds retreat into the same undergrowth, as though they were choosing a place to die. But now it was back on the shore. Dumbly, it declined to accept its fate. For a moment I thought that somehow its wing had repaired itself, miraculously snapping back into place like a dislocated shoulder. But it hadn't, and it was clear that I had to do something.

I crept up on the gull and cornered it against the sea wall, then threw my swimming shorts over its head and grabbed its flapping body. Like the swan, it made surpris-ingly little resistance, only a pathetic attempt to peck my fingers.

It was odd to see something so familiar at such close quarters: the slim elegance of its beak, long and crimson and curved to a point; the sharpness of its dusty black-brown hood defined against the whiteness of its body. It was one of the most common wild animals around, yet close to, it appeared a miracle of perfection; a perfection now irrevocably marred by the snapped bones I could feel as I examined its wing. It would never fly again, although its beady eyes looked up to the sky.

I unzipped my backpack, slipped the gull inside, and zipped it back up.

I couldn't take it home. The ride would take too long, and I feared for the bird's well-being in my bag. I cycled to the nearby country park. As I rode down the village street, I could feel the gull moving about on my back. Every now and again, it would let out a feeble squawk. Concerned it might suffocate in its nylon pouch, I stopped to unzip the bag a little, half wondering, half hoping that it might have passed away. But the faint scrabbling movements told me it hadn't given up yet.

At the park office, there was little to be done. No one was interested in the bird's plight. Someone said I was 'too soft' and that gulls were 'ten-a-penny'; a rarer animal might be worth saving, not this beach rat.

I wanted to hold up its head and show them its beautiful beak, as if it might sing its own defence. But the bird just lay there, helplessly. We were each as useless as each other. I rang the RSPCA, and was told to take it to a vet, but the journey there would just mean more stress, for us both, only delaying the inevitable. My friend, a park ranger, was working in the nearby yard.

I unzipped the bag for the last time. Richard reached in, tenderly gathering up the bird in his big brown hands as its life passed from my hands to his. 'It's been shot by an air gun,' he said quietly, then took it around the corner to wring its neck.

As I rode home, it seemed every gull on the beach turned its back to me, resolutely looking away. I imagined their reprimands as I passed, muttering 'Murderer'; they too were once persecuted and eaten, and their eggs gathered in their thousands. When I unzipped my bag again, back on the shore, I found a slick of slimy guano, and the stain of brown blood on my shorts. A single piece of down lay at the bottom. The wind whisked it out of my bag and into the waves.

I pushed out, among rafts of floating green weed, watching the ferries pass each other way off shore. In the mid-distance, a frenzy of gulls fought over an agreed, invisible point, feeding greedily on what lay below.

After a storm, when the waves roll in as if exhausted, the sea spits out strange things: huge lengths of wood which could be railway sleepers or bulwarks from ancient ships; cupboard doors and plastic seats; snaking bristles of indeterminate origin; entangled ropes covered with weed. Sometimes the scene resembles the aftermath of a battle: the unaccountable head and neck of a herring gull, floppy like a glove puppet. Bright shop-bought flowers, commemorating some unknown loss. An empty box which once contained human ashes. Above the wrack line, the charred remains of a bonfire smoulder, ringed with empty lager cans.

As a boy, I used to think what a terrible punishment it would be to have to count every pebble on the shore. Or what it would be like to lose something precious there and never be able to find it. Now, every day, I look for a stone with a hole in it. I align it to a particular point as a viewfinder; the light bursts through like a little sun, the world seen through a prehistoric telescope. They're powerful talismans, these holy or hag stones. Back home they tumble out of my pockets and over shelves and window-sills as calendars of my days; my grandmother, who lived on the edge of the New Forest, kept one in her glass cabinet, next to the rows of china spaniels.

In another glass case, in the Pitt-Rivers Museum in Oxford, there are similar stones collected from other beaches, each with a handwritten label. William Twizel, a Victorian fisherman of Newbiggin-by-the-Sea, Northumberland, arranged them around the doors of his house, an echo of his inheritance and of a people whose blue eyes were said to be the colour of the sea in front of their cottages. Next to William's stone is another example collected from Augustus Pitt-Rivers' Wiltshire estate, where it was fixed to the beam of a cottage to keep witches away.

Lieutenant-General Augustus Henry Lane Fox Pitt-Rivers – whose full name sounds more like a geographic location – was a Crimean veteran, a Darwinist, and a pioneer of British archaeology. He ordered his collections according to type and use, rather than date and provenance. Now they lie crammed in table-top and wall-cases, with fossils and fans, fetishes and tribal masks all jumbled together in a dark galleried hall resembling a particularly gloomy department store.

Some stones come from Craig and Ballymena in Northern Ireland or Carnac in Brittany, and were used to protect cows, tied to their tethering stake or between their horns to prevent pixies stealing their milk. Horses, too, were hung with hag stones to stop witches riding them during the night, and on Dartmoor stones were worn around human necks or nailed to bed-posts to defend against nocturnal demons. The seventeenth-century Brahan Seer of the Scottish Highlands was said to have had such a stone through which he could see into the other world, although he was burned in a spiked barrel of tar on the beach at Chanonry Point for his trouble.

In a 1906 essay on 'Witched Fishing Boats in Dorset', Dr Henry Colley March observed that Dorset fishermen would tie holy stones to the bows and thread their start-ropes through them; it was also the custom in the same county to attach the key of a house to a holed beach-stone for luck, he noted. The learned Dr March went on – in the antiquarian tradition of his peers such as Pitt-Rivers, M.R. James and James Frazer – to discern an association with the megaliths of ritual land scapes in southern England, Orkney and Brittany; half-natural, half-constructed objects charged with the power of the past. 'It is impossible not to see a like motive for the ancient practice of dragging a sick or epileptic child through a hole in a large "druidical" stone.' They might as well be threaded with sacrifices to unknown gods, although to my eyes they resemble little modernist sculptures.

But you can't keep a beach in your house, despite the teetering piles on my shelves, a terminal moraine of memory, accumulating dust composed of the decomposing me. Deformed oyster shells grown around odd pebbles, smooth grey wooden shapes with knots for eyes, sand-blasted shards of Victorian glass, chunks of marmalade pots, soft stems of clay pipes once sucked between moustached lips and discarded like cigarette butts, and bits of blue willow-pattern plates awaiting ceramic resurrection; all of them tumbled together by the tides. As a boy I listened to the rushing noise inside shells; the same sound has been in my ears for years now, a perpetual ocean in my head.

In Iris Murdoch's strange novel *The Sea, The Sea*, published in 1978, the celebrated actor Charles Arrowby retreats from London to write his memoirs. He lives alone in a ramshackle house on an unidentified English coast, having come to the sea in search of 'monastic mysticism'. The events that follow – a series of impossible coincidences and fantastical happenings played out in a Shakespearean manner, with Arrowby as a kind of Prospero – take place over one summer. They occur in an indefinably apocalyptic time and place, as if society were on the verge of disintegration – as indeed it seemed to be at the time – although ostensibly all is normal in the world beyond Arrowby's coastal retreat.

As he relishes his solitude and the splendour of its setting – swimming off the rocky shore, ordered by its tides or the way he gets in and out of the water via a rope – Arrowby's idyll is broken one morning by what he imagines or perhaps actually sees: a maned and toothed leviathan. '*I saw a monster rising from the waves,*' twenty or thirty feet high, coiled and spiny, its head crested and with sharp teeth and a pink mouth. '*I could see the sky through the coils.*' Although Arrowby puts this terrifying vision down to an acid flashback, a legacy of his misspent youth, the beast is an omen of all that happens afterwards as his former lovers and enemies come back to haunt him. Nor was it coincidence that this unsettling apparition is an echo of a scene in Racine's *Phaedra* in which a sea monster appears, so fearsome that it infects the air and causes Hippolytus' horses to drag their master to his watery doom.

Throughout Murdoch's tragi-comic drama, the ever-changing sea abides, a character in itself, a mindful reminder of her hero's impotence to change his fate and manipulate his friends, despite his deluded attempts to do so. He seems locked into what is happening, all the while documenting each oddly concocted meal, even at moments of great crisis. He eats: tinned macaroni cheese 'jazzed up' with cold courgettes, 'Battenburg roll and prunes', 'boiled onions served with bran', 'poached egg on nettles', 'a little cold jellied consommé straight out of the tin', all washed down with Spanish wine bought from the nearby Raven Hotel. Such details serve only to make the story more bizarre. Having abducted the woman whom he had loved as a boy, and who, by extraordinary circumstance, he suddenly discovers to be living nearby, the book ends with Arrowby – who has nearly drowned, violently, in the sea in typically mysterious circumstances – experiencing an epiphany in the shape of four seals bobbing in the water, 'their wet doggy faces looking curiously upward... And as I watched their play, I could not doubt that they were beneficent beings come to visit me and bless me.'

Murdoch took her title from the cry of the Greek warriors who finally saw the Black Sea after fighting against the Persian empire, a sight that heralded home. As a writer, she was criticised for her apparent belief in myth and monsters: in person, as in her work, her fierce intelligence contrasted with a faint naïveté. She ended her life losing her senses in public, suffering from dementia and yet being taken everywhere by her writer husband. I'd often see her at literary launches: a ghostly, silver-haired figure with flickering eyes and a fixed smile, lost in a corner of a room that might have been any room, anywhere, with anyone in it.

The sea sustains and threatens us, but it is also where we came from. Some consider that the relationship is closer than we think. Callum Roberts, among other scientists, has noted that the ratio of subcutaneous fat in humans is ten times that of other primates, nearer to that of a fin whale. From an evolutionary point of view, such human blubber would make little sense for a land hunter, but it would be eminently useful for an 'aquatic ape' which developed by the sea. Equally, we cannot fly or even run as fast as other animals, and we lack hair to keep our bodies warm, but we can swim and dive – skills which would not make sense, some say, unless we were made for or at least shaped by the water.

First proposed by Desmond Morris and subsequently explored by Elaine Morgan – who saw a certain prejudice in the way in which her ideas were rejected – the 'aquatic ape' theory is controversial, dismissed by scientists suspicious of its simplicity. Perhaps there is something a little too perfect about the notion that rather than descending from the trees to hunt on the savannah, we gravitated instead to the shore, not least because it argues against the idea that we are defined by our ability to kill. Yet new evidence suggests that a diet sourced from the ocean may have provided the fatty acids that enabled our brains to grow, and that we stood on two legs to wade as we scavenged for shellfish on the shores of our earliest home in sub-Saharan Africa. That we were, and are, intimately linked to the sea.

Other factors have been marshalled to support Morgan's argument: that we are prone to dehydration in a manner which would not be helpful to savannah-dwelling animals, and that we exhibit an instinctual breath-holding reaction when we plunge into water: other terrestrial mammals cannot regulate their reflex breathing. Does this mean that we were once well used to entering the sea, perhaps sticking our heads in to search for food, or even spending longer periods there? The Belgian anthropologist Marc Verhagen and his colleagues believe it is possible, arguing that our wide shoulders are more suited to swimming than running, and that we might owe our long legs and long strides to forebears who foraged in the shallows.

Our vestigially webbed fingers have also been claimed as an amphibious refinement of this watery life – just as the seabird hunters of St Kilda developed broad feet from climbing the cliffs, and the sea gypsies of south-east Asia, used to diving in the shallow seas, have eyes that appear to focus as well, if not better, underwater. Even our organs contain a memory of the sea. Our kidneys evolved to deal with excess salt, to which our evolutionary ancestors were subject; being fifty per cent water, we all contain the sea inside us.

While many scientists dismiss the notion of an aquatic ape, the proposal is intriguing: that we owe our development and our dominion, our intelligence and even our souls, to the water, although we live out our lives on the land.

We cannot resist; we are all watergazers. And like so many others, my own, more recent ancestors also felt the urge to travel over the sea, in search of something new.

My great-great-great-uncle, James, the eldest of William Nind's thirteen children (of whom at least three died in childhood), was born in Ashchurch, a hamlet near Tredington in Gloucestershire, in 1782. The Ninds had been farmers since at least the sixteenth century, living in the same small triangle of fertile land at the foot of the Cotswold hills, moving between villages such as Beckford, Walton Cardiff, Ashchurch and Alstone, near the towns of Tewkesbury and Cheltenham.

As the eldest son, James was expected to follow his father, working the land. But in the family tree she compiled in her fine handwriting, my aunt recorded the tradition that James left England for Ceylon to acquire a plantation (probably of coffee, rather than tea, which was not grown on the island until the mid-nineteenth century). He was also said to have been involved in another trade: human trafficking. Slavery had been banned by Britain in 1807, but continued in its colonies until 1838, and James was engaged in 'blackbirding', a kind of kidnap in which native people were lured onto ships and abducted, to be sold on as indentured labourers. As a result of his activities, James Nind amassed a fortune worth three hundred thousand pounds, only to die at sea, sailing from Ceylon on a ship called *Breezy Horse*.

Ever since I heard about it as a boy, I've imagined that scene: a storm-tossed vessel, my ancestor tipping over the gunwales in a flash of lightning. Or perhaps he went of his own accord, like Captain William Ostler of the *Marquis of Hastings*, who, on his way back from New South Wales and China in September 1827, 'threw himself overboard in a fit of insanity off the Cape of Good Hope, on the night of the 9th September. A paper, containing the following words, was found lying on the table of his cabin in the morning: "A bad crew and bad chiefmate is the destruction of William Ostler."' Whatever the manner of James Nind's death, it was certainly unforeseen: he left no will, and his relatives were unable to claim his money. A moral return, perhaps, for such immoral gains.

For years afterwards the Ninds tried to find out what had happened to James, and his fortune; there were no records of his estate, his ship, or his death to be found. Yet the rumours persisted, fed by the prospect of lost riches. In 1921, a syndicated story appeared in the American press under the headline 'Unsolved Mysteries'. It speculated that James Nind had actually come to America, rather than Ceylon, along with his brother William, and had accrued his wealth in South America under an assumed name. 'The theory is that James Nind after living in New York for some years went to one of the South America Republics...building up a fortune as so many adventurers from the Anglo-Saxon race have done in different parts of the world. In these countries the state of society is so unsettled that many obstacles might be thrown in the way of recovering the Nind fortune.'

None of this makes much sense – the report, in the *Galveston Daily News* of Texas, confuses more than it reveals – but the fact that it appeared in a newspaper gave credence to the story. With rumours of impostors turning up in Cotswold villages, only to disappear again, there were even hints of conspiracy: 'It seems clear that someone is interested in the matter aside from the Nind heirs...' Indeed, this family intrigue, with its echoes of a novel by Conrad or Dickens, would be strangely replayed in the next generation.

James's nephew, also named James – my great-great-grandfather – was born one of nine children, to Isaac Nind, a gentleman farmer in Tredington, in 1824. As the eldest surviving son, he stood to inherit substantial property: his father owned two hundred acres and employed four labourers, as well as six domestic servants. Yet in 1850, aged twenty-five, this fair-haired, handsome young man followed in his uncle's wake and left England. In his case, he definitely sailed to America, apparently drawn there by the promise of a better life.

But James had another reason to leave home. That summer, Sophia Clarke, a twenty-one-year-old woman from the neighbouring village of Gotherington, gave birth to their illegitimate daughter Rosa. The records claim, somewhat mysteriously, that Rosa was born at sea. Had Sophia travelled to America with James, and decided to return? When asked about her grandfather, my grandmother would only say, 'There was a young man who was sent to America to make a fresh start.'

Although James already had family living in the United States – his aunts Dorcas and Judith had emigrated there some years before – it must have been an extraordinary contrast, to leave the lush confines of the Cotswolds for the vast and still largely unexplored continent. That may be why he settled in Lowville, in the foothills of the Adirondacks, New York State, a reminder of home: good farming country, like its neighbouring state, Massachusetts, where, in the year that James arrived, Melville was writing *Moby-Dick*. By 1859, when he was visited by his sister Mary Ann and her husband John Freeman, James was married with twin boys. But that year he announced his intention to go west to the goldfields of California, lured by reports such as one in a Buffalo newspaper which claimed that prospectors could turn fifty dollars into five thousand within twelve months. Mary Ann, initially keen to join him, decided not to go as she was about to give birth.

She had a lucky escape. James and his family were last heard of in Davenport, Iowa, from where they joined the wagon trail. From the 1840s to the 1860s, four hundred thousand travelled west, an unprecedented exodus of people from all around the world to the remote Pacific coast: Mormons, miners, farmers and families in search of fortunes or religious freedom or any kind of new life. The journey would take half a year and was fraught with danger. Wagons were towed by oxen across plains as yet unclaimed from Native Americans, through the desolate landscapes of the Great Salt Desert and over the mountain range where the Donner Party had resorted to cannibalism in their despair. This mass migration had its own power to alter the environment, not least in the hunting of bison, about to be driven to the verge of extinction.

Did James and his family make it as far as the Great Plains, travelling by prairie schooner, sailing through endless seas of grass? I once visited those same fields, without knowing that my ancestor might have passed that way. It was as far from the ocean as I've ever been, and I remember swimming in an open-air public pool on the outskirts of Red Cloud, Nebraska. It looked like a little piece of the sky fallen to earth. All I do know is that James wrote a letter to his sister, Mary Ann, sent back east, although it survives only in her report. 'A wagon train can pass through the grassland seas,' she wrote, 'they had circled their wagons to camp and put the boys under the wagon.' There, in an extraordinary, unbelievable stroke of bad luck, the boys were both bitten by a snake, and died. James also reported that his wife was ill. And that was all; except for his last words, left hanging in the air: 'I don't know.'

James never reached his destination. Perhaps he and his wife succumbed to disease. Cholera was rife among the migrants, 'the destroyer...let loose upon our camp', as one settler wrote. Or perhaps, as family tradition suggests, he was killed by Indians. It is not an entirely fanciful notion: such attacks were the second most common cause of death for the travellers moving in great numbers through Native American territories. James's sandy hair would have made a fine scalp.

I can't quite believe myself descended from these romantic ancestors, or imagine what they experienced, or inflicted. Their stories are beyond the reach of the brown-grey ghosts of the family album. They happened before the casual snaps in trellised gardens and on seaside promenades, and they suggest more than they tell. James's sister Mary Ann, and their brother William, who followed them to America, would lead quieter lives, settling in small towns on the shores of Lake Erie, south of Buffalo and Niagara Falls. Yet for them too, leaving England was an adventure: Mary Ann would recall that on the voyage out from Liverpool, the ship on which she was sailing passed another vessel on fire, but their captain did not stop, although the law of the sea demanded that they should.

Back in Gloucestershire, Sophia went on to marry twice, each time to men from her own parish. A single image of her survives: a tintype photograph, corrupted with age, showing her in a patterned dress. Her face is strong, her cheekbones high, her stance determined. She looks like my mother, who had the same Titian red hair; it is not hard to read in her eyes what she had lived through. Sophia brought up her daughter, assisted by the Ninds, who acknowledged her as one of their own. In her teens,

Rosa became a nursemaid to a naval family in Plymouth, before marrying and having her own family, among them my grandmother. She looks quite proper in her crinoline. But in every census record in which she appears, until her death in 1920, one year before my mother was born, she continued to state that she was born at sea, as if to obscure the shame of her illegitimacy.

My father's family also crossed the sea; like my maternal ancestors, they too were caught up in an age of mass migration. My great-grandfather Patrick, named after the saint who had converted Ireland and driven the snakes from its shores, was born in Blanchardstown, a village outside Dublin, in 1856. The island was still suffering the after-effects of the Great Famine, a good enough reason for his departure for Liverpool sometime in the 1870s. Settling in Lichfield, he married an English girl, a servant in the same household in which he worked as a coachman. The pair then moved to the former whaling port of Whitby, where my grandfather was born in 1885, on a street at the end of which, in the previous century, James Cook's *Endeavour* had been built.

His eldest son, my father, a dark-haired, good-looking young man, left the depressed streets of the north for a new life in Southampton in the 1930s. He had been born in the model mill town of Saltaire in 1915, but was brought up in Bradford. His journey south was the equivalent of the Ninds' voyages, the result of greater events, of disaster and opportunity. Later he'd speak of the deprivation he had witnessed in his home town, of starving families fighting over food, of rats running down the street, and of a man found hanging in an outhouse on nearby wasteland.

Perhaps that's why the rest of my father's life was so resolutely ordinary and ordered. He worked for the same cable company for forty years in a redbrick factory built on reclaimed land between the docks and the walls of the old town, a hundred yards from the station where he had first arrived. Every day, at the same time, my mother waved him off. Every day, at the same time, he came back for tea. He might as well have been clocking in and out of his own home. What he did between his punctual departure and his prompt return was a mystery. He seldom spoke about his work, nor did we ask him about it.

In the summer after leaving school, I went to work in the same factory. I was a fitter's mate, a position which required me to wear grey overalls and accompany my designated fitter on various jobs, in none of which did I perform any kind of useful function. As we set out for the day or came back to the workshop before going home, having smeared our clothes with grime to make it look as though we'd been busy, I'd look up into the glass box where my father worked above the factory floor and see him there with his colleagues. They wore bright white coats to distinguish themselves from us in our oily overalls, and they all seemed to wear spectacles, of a National Health design; anything else would have been an unwonted vanity.

Lit by the luminous shed over our heads, for a few, tantalising glimpses I saw my father as he was seen by others: as a person, rather than a dad, up there in the Test Department, with its graphs and dials and meters of resistance, dedicated to certifying six-inch cables which would be wound onto elephantine wooden drums, stencilled with yellow paint and unrolled under the ocean, anchoring England to America.

Now I see my father through the framed photographs that still stand in my mother's bedroom, and I'm taken aback to realise how much like him I've become: from his shorts and the bag slung about his neck as he stands on a seaside cliff, to his love of the sea itself, which he would scan through his heavy binoculars, breathing in deeply as if to clear his lungs of Bradford soot. We share the same shape, the same bones that show through my skin; what I have been and what I may become. How could I have thought myself so different?

Recently I went back to Yorkshire. It was a long journey, each connection with its own story. The dawn suburban train, with workers' eyes stuck with sleepy dust and toothpaste masking morning breath. The peak commuter train, full of furious notes on laptops and phones forever seeking attention. The midday train, relaxed with senior citizens, and students locked into their own, dreamier electronic worlds. And finally an afternoon shuttle explosive with children turned out of school, shouting and laughing and dodging the ticket collector as he came down the compartment, jumping off at the next halt, spilling onto the platform and vanishing as quickly as they'd appeared. As we were drawn to the north, to its wasteland and wealth, I was struck, as I watched through the window, how different it was, an abrupt view of smoke stacks, as evocative as any university's spires to its alumni. For a few seconds I saw the town set in its sulphurous dip, its name taking me back to childhood visits – *Bradford Interchange* – as I sat in the modern carriage forty years later, listening to languages my father never heard.

I might have been the same person as him. As a child I never saw different people – unless in the mirror, when I dressed up as a Red Indian. Our family never went abroad; I travelled only in books, although I was born in a port. Our first fear is abandonment; our last, too. We all leave home to find home, at the risk of being forever lost.

2
The white sea

A raven alights
At God's ear.
Tidings he brings
Of the battlefield.

W.G. Sebald, 'Time Signal at Twelve'

F rom the far side of the Solent, I look down on the slow-moving sea, four hundred feet below. At this altitude the water is abstract, no longer audible. The sheer chalk cliffs on which I stand are a dazzling end to England. They're almost too bright to look at, bouncing the light up and down. Their reflected whiteness illuminates the sea from beneath, in the way Pre-Raphaelite canvases were underlit with white lead paint. And like one of those hyper-real scenes, this is a heady view. Every detail is heightened, as if it all weren't quite believable. The grass is sown with wild flowers, pink, blue and yellow. Everything seems bleached and saturated at the same time. Pale-bodied gulls ride level with my eyes, studiously ignoring me.

I wasn't meant to be here today. Playing truant on a weekday, I put my bike on the train, next to the others in the cycle compartment, a confraternity of machines grungy with oily chains and mud-spattered frames. Unfolding my much-creased map, I planned my ride to the sea – only to be informed of engineering works. Wondering what to do,

I overheard a group of young German students discussing their trip to the Isle of Wight. So I bought the same ticket as them. At Brockenhurst I switched trains for the short ride through the forest, the bracken billowing by the side of the tracks. Ahead lay Lymington pier, and the ferry ready to cross the strait.

On a fine spring morning, it was busy with two-way traffic: expensive leisure craft setting out for the Solent; the first swallows of the season, returning from Africa. As the boat nudged out into the channel, I looked down on the salt marshes on either side. The birds were caught up in spring fever, feeding and courting. There was a sense of expectation, as if the day had opened up after a long, closed-down winter. I remembered childhood visits to the island, when my dad would joke about having our passports ready for the ferry ride. We'd stay in a railway carriage converted into a kind of chalet, set in a farmer's field outside Cowes. It was like living in a tin can for two weeks, with narrow compartments for bedrooms, and a low curved ceiling. At night, moths gathered round the fragile gas mantles, and bats fluttered outside in the sort of darkness we never saw in suburbia.

There are many speculations about the island's name. The Old English *wīht* may relate to the Germanic root, 'little spirit' or 'little daughter'; it may also refer to 'that which is raised over the sea' – an echo of the island's Roman incarnation as Vectis, or lever. Although none of these is certain, such names seem to express the physical separation of this fragment which rises out of the sea, the geological end – or beginning – of the soft spine of calcium carbonate which twists through England. The island was formed when the river Solent rose after the Ice Age; from its sea bed the bones of aurochs and woolly mammoths have been recovered, along with the stumps of petrified forests. Now the island stands between the mainland and the Continent as a last vestige of Englishness; in his eccentric essays on *Desert Islands*, Walter de la Mare saw it as a doormat 'very dear to the eyes of an Englishman on his way home'. Although its megafauna have been replaced by cows and its megaliths by bungalows, its gentility is a mere veneer. Like all islands, this is a place defined by feral forces.

At Lymington, on the mainland, the island is at its closest, only three miles distant, although that narrowness causes the water to surge in a race enough to intimidate the most experienced sailor. From my blustery vantage point on the top deck of the ferry, where the wind whips my hood and forces me into fingerless gloves, I watch the western end of the island approach. Here it rises highest as it enters its final act of disintegration, sending bits of itself out to sea, rocky icebergs calved from a glacier of chalk.

As the ramp hits the hard at Yarmouth, I'm let loose like a sheep from a truck. I ride along the low river valley that almost entirely divides the extreme, prow-shaped tip of West Wight from the rest of the island. My 1950 Ward Lock guidebook – bound in red cloth and filled with airy advertisements for a place that has barely changed in sixty years – tells me that 'in stormy weather the sea has been seen to break over the narrow ridge of separation and mingle its salt waves with the fresh waters of the riverhead'. Far from seeking to bolster this slender bar, the islanders sought to increase the gap. In an earlier guide, published in 1856, William Davenport Adams, a former teacher, noted that the strand of shingle was 'formerly... much less; so that the inhabitants of the island proposed in the reign of Edward I, to cut through the isthmus, and thus to form for themselves an almost impregnable retreat, when the island was invaded by hostile bands'. This insular wedge of land even aspired to its own status as the Isle of Freshwater, an island-within-an-island. But then, the Isle of Wight itself has always resisted attempts to link it to the rest of England; having long loosed its moorings, it prefers to float free.

I swim off Freshwater Bay, overlooked by the same Albion Hotel from where Adams began his tour. 'Though the beach is pebbly and rocky, bathing is good, the sea being in calm weather remarkably clear,' my 1950 guide informs me. 'Boating under ordinary conditions is quite safe, but for trips of any length, a man who knows the coast should certainly be taken.' The water is salty and buoyant. I'd been looking forward to it and dreading it at the same time. There's no comforting land on the horizon, only the steely sea. Around me old plastic containers bob, tethered to lobster pots below. The cliffs rise as a backdrop, their soft chalk embedded with flints like nuts in a bar of nougat. Grey and white pebbles as big as tennis balls roll under my feet. The water is cold and my swim is brief, although it has the usual effect of washing my sins away.

I'm warming my hands on my flask of tea and thinking about leaving when the German students arrive. One of them leaps onto the promenade railing and, in an acrobatic defiance of gravity, holds his body vertically for a split-second, supporting himself on one hand. Another changes into brash shorts and black goggles and strides into the water. He swims straight ahead, disregarding the weedy rocks. Having reached his appointed end – his route might as well have been marked out in the lanes of a public pool – he swims back to his friends' applause.

After these breezy exertions the students pack up and go. Emboldened, I plunge back in, following the channel the boy had pioneered. As I swim, still wary of the rocks, a black-painted ketch sails into the cove and straight at me. Its crew comprises a long-haired man – perhaps one of those useful locals recommended by my guide – and a younger woman who leans over the side and shouts at me, 'Are you mad?'

Back on the beach, she produces a blue plastic bucket. Inside is a slithery selection of still-twitching fish: whiting, catfish, and other brown-bodied, slimy species.

Warming to her audience – a man and his young son have come up to peer into the bucket – she says that the sea is two degrees warmer than it should be at this time of year, that fish are appearing which ought not to be here. I ask if they ever see any dolphins out there. She defers to her captain, who says yes, a few miles out.

'The *Common Porpoise*,' Mr Adams reassures me, 'occasionally passes along our southern shore in small shoals.' But even bigger cetaceans have been found here. In 1758, a Royal Navy ship captured a sixty-six-foot fin whale which had been seen floating dead in the water, only to leave go of it off the Needles, from where it washed up at what would become known as Whale Chine. It took a century for another fin whale to appear at the feet of Tennyson Down, in 1842; it ended up in the amusement park at nearby Blackgang Chine, where its enclosed bones became a kind of cavernous shopping-bazaar-cum-boatshed, lined with plates, baskets, pictures, shells and glass lighthouses filled

with coloured sand from Alum Bay, while its empty jaws pointed forlornly to the window and the sea beyond. It was still there when I visited the island as a boy, although I remember being more excited by the plaster gnomes and toadstools outside.

Leaving the beach, I turn inland, passing a pair of bay-fronted villas at the foot of the downs – the sort of seaside houses that offer bed and breakfast, polyester chintz and cracked shower cubicles; but a crenellated tower and oriel windows suggest another story.

Dimbola Lodge was the home of Julia Margaret Cameron, who, swathed in purple paisley, her hands stained 'as black as an Ethiopian Queen's' by silver nitrate, created alchemical images in her greenhouse, having summoned subjects spotted in the streets from her eyrie. As the name of her house suggests, she too was a migrant, an exotic bird blown onto the island's shores.

Cameron was born in Calcutta in 1815. Her family name, Pattle – close to the Hindu 'Patel' – indicates an Asian ancestry, evident in her broad forehead and 'Pondicherry eyes', the same features seen in her great-nieces, Vanessa Bell and Virginia Woolf. Theirs was a family which looked east. Julia's husband Charles, twenty years her senior, had fallen in love with Ceylon as a young man, and continually returned there. Their house in Freshwater was named after the family's coffee plantation, while the subcontinent's influence was reflected in its Indian gothic arches.

Writing from Dimbola, Julia tried to entice Charles back with her fanciful comparisons. 'This island might equal your island now for richness of effects,' she declared, as if West Wight had somehow magically moved to the Indian Ocean, or at least been towed there in her imagination. 'The downs are covered with golden gorse & beneath them the blue hyacinth is so thickly opened that the valleys look as if the "sky were upbreaking thro' the Earth"…The trees too are luxuriant here – far more flourishing than they usually are by the sea – and Alfred Tennyson's wood may satisfy any forester.'

Here Cameron pursued her art. Her photographic images show only human faces, but they're lit by the chalk downs and the limitless sea around them. With their shifting focus and depth, their draped shawls and tresses of hair, they resemble underwater scenes. They are worlds of their own, more organic scenes than portraits. On her way south from London, Julia would change at Brockenhurst, where her photographs of Darwin, Browning, Tennyson and others still hang in the booking hall, along with a hand-written sepia inscription:

> This gallery of the great men of our age is presented for this Room by Mrs Cameron in grateful memory of this being the spot where she first met one of her sons after a long absence (of four years) in Ceylon. 11th of November 1871.

She must have made for a lively travelling companion. Henry Allingham encountered her at the station, 'queenly in a carriage by herself surrounded by photographs... talking all the time. "I want to do a large photograph of Tennyson, and he objects! Says I make bags under his eyes – and Carlyle refuses to give me a sitting, he says it's a kind of *Inferno*! The *greatest* men of the age (with strong emphasis), Sir John Herschel, Henry Taylor, Watts, say I have immortalised them – and these other men object!! What is one to do – Hm?' (Her conversation was constantly punctuated with 'eh?' and 'hm?').

Julia presided over an unlikely irruption of bohemianism on this islet. 'Everybody is either a genius or a painter or peculiar in some way,' complained one visitor, 'is there *nobody* commonplace?' With its perpetual parties and play-acting, Freshwater was one long performance, and Dimbola its proscenium arch. Such was the throng of art and letters that one writer compared it 'to Athens in the time of Pericles, as being the place to which all the famous men of the reign of Queen Victoria gravitated'; another considered its society closer to a French salon than any English gathering.

It's hard to imagine now that this sleepy village should have been filled with such fanciful displays. Leaving Dimbola and its eccentric cast behind, I push my bike up the hill – and nearly blunder into a skylark at my feet. Ahead of me, the island rises, 'walled up from the ocean by a bulwark of immense cliffs', as Adams writes, caught up in the spirit of the sublime. 'A mighty barrier, truly! but yet not altogether impregnable against the assaults of the sea. Their glittering sides are strangely branded, as it were, by dark parallel lines of flint, that score the surface of the rock from misty ridge to spray-beaten base. Huge caverns penetrate into their recesses. Isolated rocks frown all apart in gloomy grandeur. Immense chasms yawn...'

Couched in such breathless terms, the terrain takes on the air of Julia's amateur dramatics: the celadon sea, the lime-lit cliffs, the sun switched on and off by a shifting curtain of clouds; a Victorian stage on which an over-made-up Edgar might be telling his blinded father Gloucester, 'How fearful/And dizzy 'tis to cast one's eyes so low!' This is a place used and shaped by every species that visits it. The turf is as clipped as a bowling green, the work of rabbits whose burrows riddle the soft chalk. Violets and gentians stud the grass like purple stars. There's barbarity here too, invoked by a raven perched on the edge, but it is upstaged by an altogether more diminutive bird: the wheatear, newly arrived from the other end of the earth.

I say wheatear, but that is a prudish corruption of its wonderfully common name, white arse, one which reaches back to the Old English *hwīt*, and *ærs*, its distinctive tail. Its binomial, *Oenanthe oenanthe*, is almost bigger than the bird itself, and derives from the Greek for wine flower, since its bearer's arrival in southern Europe from Africa coincides with the blossoming of the vines. Weighing barely twenty-five grams, the northern wheatear accomplishes a migration among the greatest of any songbird, round trips of up to eighteen thousand miles from South Africa to Alaska. Those that land in southern England, however, choose to nest here, using abandoned burrows.

Under the cover of a hollow created by subsidence, I crawl along to get a better look. They're exquisite birds. I focus on one male as it hops about on the precipice. Its bluish-grey back reminds me of a guardsman's greatcoat, and its pale belly is blushed with a seductive peach tint. Its face is pharaonically masked by a black eye stripe. It stands on an outcrop of chalk, proud and alert on surprisingly long legs, the light reflecting up at its breast. As it runs along, it launches into short skittering flights, flaunting its white arse. It is loyal to the downs, a welcome resting place after all those thousands of air-miles, but a fateful choice for its predecessors, which were rewarded for their efforts by being baked in a pie.

According to Gilbert White, writing in the 1770s, wheatear (as a respectable cleric, he was bound to use the bird's less indecorous name) were 'esteemed an elegant dish'. 'At the time of the wheat-harvest they begin to be taken in great numbers; are sent for sale to *Brighthelmstone* and *Tunbridge*; and appear at all the tables of the gentry that entertain with any degree of elegance.'

To supply such slender fare, shepherds left their flocks every summer – to the annoyance of their employers – setting their snares on the feast of St James, 25 July. Thomas Bewick, writing two decades after White and referring, in his no-nonsense northern text, to the White-Rump, noted that the traps were made of horse hair, and hidden beneath a piece of turf. He reckoned 'near 2000 dozen' were taken each year around Eastbourne alone, to be sold for sixpence a dozen in London, 'where they are...thought not inferior to Ortolan', a small bird which was and still is eaten in France. (A captive ortolan's eyes were poked out; it was then force-fed oats before being drowned in brandy and swallowed beak-first, its carcase covered with a napkin to hide the shameful act from the eyes of God.) In 1825, Mrs Mary Trimmer, 'author of *Man in a Savage and Civilized State*', updated Bewick's book, silently removing 'White-Rump' from its chapter heading, adding that, 'as they are very timid birds, the motion of even a cloud... will immediately drive them into the traps', and noting that a single shepherd could catch eighty-four dozen birds a day.

As the overfed inhabitants of the watering holes of Brighton and Tunbridge Wells, themselves as migratory as the birds, sucked and chewed one white arse after another, it is little wonder that, as the Reverend White noted, 'About *Michelmas* they retire and are seen no more until *March.*' White, though, had no idea where they went.

4   The Wheatear

They may be scared by clouds, but wheatear are bold birds, hopping close to my feet. To Derek Jarman, walking in Dungeness where he had retreated to his tar-black Prospect Cottage, and where he gathered holy stones for friends to wear around their necks as though to protect them from the nearby nuclear reactor, 'the snobbish wheatears complain – *tch! tch!*' But Tennyson heard them differently –

> *The wheatears whisper to each other:*
> *What is it they say? What do they there?*

– as he pounded the island's downs in his military cape and broad-brimmed hat and oath-catching beard.

Tennyson was born in Lincolnshire in 1809. 'Before I could read,' he remembered, 'I was in the habit on a stormy day of spreading my arms to the wind, and crying out "I hear a voice that's speaking in the wind."' He was a poet attuned to nature in what we consider to have been an age of science and industry. Yet the era itself was wild and prophetic, as filled with female Christs and millenarians as it was with evolutionists and industrialists; as much concerned with myth and mortality and looking backwards as it was with moving into the future.

Tennyson's most famous work, *In Memoriam*, was a tribute to his handsome young friend Arthur Hallam, for whom he had 'a profound affection' and who died of a cerebral haemorrhage at the age of twenty-two:

> *Forgive my grief for one removed,*
> *Thy creature, whom I found so fair.*

When the poem was published in 1850, many readers thought these lines must have been written by a woman. To Tennyson, Hallam had become conflated with the young hero of Thomas Malory's *Morte d'Arthur*, the inspiration for his own *Idylls of the King*, which in turn Julia Margaret Cameron re-imagined in her dreamy photographs. Their Isle of Wight was a new Albion, clad in its white cliffs, as if its downs and beaches were a concentration of all that Britain represented. After all, the entire empire was overseen from this sliver of southern England, floating in the Channel like Lyonesse, the mythic isle where Arthur was said to have died. A few miles from Freshwater, in her island fastness of Osborne, Victoria, who would lose her own prince to typhoid, pronounced the poem to be her favourite work of literature after the Bible.

Yet Tennyson's romantic sentiments were not quite as they seemed. His words expressed the overturning of old certainties in the age of Darwin's discoveries:

> *Tho' Nature, red in tooth and claw*
> *With ravine, shriek'd against his creed.*

The way that nature was seen was being overturned. In 1844, Robert Chambers' *Vestiges of the Natural History of Creation* was published anonymously, and sensationally, addressing the idea of the transmutation of species. It was avidly received, not least by Tennyson, who acknowledged that it explored theories he himself had addressed in his work, and by Prince Albert, who read it aloud to the Queen. As Tennyson was drifting into his Arthurian *Idylls*, Darwin was writing *On the Origin of Species* in a hotel on the same island. It was as if the world were being reordered here, at England's outer limits.

'The far future has been my world always,' Tennyson said. He retreated to Farringford, his house in Freshwater, and retitled his portrait by Julia Margaret Cameron 'the dirty monk'. But he soon realised that nature could not protect him; that despite the woods around his house and the buttresses of the sea and its raven-haunted ravines, he was also surrounded by his words, as ensnared as a wheatear, unable to escape the place he had assumed in the imperial pantheon. By 1869 the sightseers had become too much, and Tennyson left Farringford for Surrey, where, twenty years later, he would receive the news that his son had died at sea on his way back from India.

Alfred Tennyson.
from the photograph by Mrs Julia Margaret Cameron.

By then his former neighbour had long since decamped. After years persuading ungrateful great men to pose for the plate-glass negatives which would indeed immortalise them, Julia and her ailing husband sailed from Southampton on the P&O ship *Mirzapore* in the autumn of 1875, bound for Ceylon. With them they took their coffins, packed with their possessions, knowing they would never return.

On the island's western coast at Kalutura they set up home, covering their bungalow's walls 'with magnificent photographs' as a visitor, Marianne North, recorded; 'others were tumbling about the tables, chairs, and floors with quantities of damp books, all untidy and picturesque; the lady herself with a lace veil on her head and flowing draperies. Her oddities were most refreshing.' Julia photographed local people in the same style in which she had posed the great and good of British society. But her new island idyll – as if she had finally accomplished the feat of towing the Isle of Freshwater to the Indian Ocean – was only fleeting. Four years later, having moved to a house high in the tea-growing hills of Nurawa Eliya, then known as Little England, Julia died, looking up at the stars, uttering her last word, 'Beautiful.'

As I resume my climb, I'm forced to stop every now and again – not because I'm running out of breath, but because the view itself is breathtaking. Behind me is the landscape I have known all my life, levered up and flattened at the same time; a three-dimensional map of my hinterland, from Bournemouth's beaches to Portsmouth's towers, signalling the south-eastern sprawl towards London.

Everything falls away, as if seen from a lurching helicopter. At the feet of the cliff, the sea is turned cloudy, as though scattered with bath salts. Beyond, its open surface is pooled with shifting shafts of sunlight and mercurial upwellings. It seems to move under itself, inexorably, a great slow mass beneath a rippling skin. The quiet is unnerving. You'd think such a vast vista would evoke an awful noise, echoing with what Tennyson called 'the moanings of the homeless sea'. This is as far as I can go. Any further and I'd have to take to the water. I feel a sudden surge of inexplicable homesickness, perhaps because I can see my home, near and far away.

I think the silence is getting to me.

William Davenport Adams bellows – it's the former schoolmaster in him. 'Oh, there is such a glorious prospect from this bold headland!' he shouts, making himself heard over the long-departed wind, 'a vast sweep of the Channel blends strangely in the distance with the deep-blue sky, and when you turn, lo, beneath you lies the silent peninsula-plain, with its villages, and wooded knolls, and sequestered farm-houses, and sparkling streams.'

I've reached the top, staked out by a grandiose granite monument, erected in 1897 –

IN MEMORY

OF ALFRED

LORD TENNYSON

THIS CROSS IS

RAISED A BEACON

TO SAILORS BY

THE PEOPLE OF

FRESHWATER &

OTHER FRIENDS

IN ENGLAND

AND AMERICA

– while the etched aluminium plate of a range-finder at its feet tells me that if I sailed due south-west from here, I'd reach the Azores. The world below is dazed by the light, as if everything had evaporated into the ether.

Except that someone has neatly used chalk fragments, all of a size, to spell out the word *Google* on the grass. Next time I come by, the word will be replaced by a pair of lovers' names.

The clouds are rushing under the sun, casting shadows on the sea, creating fluffy stealth bombers. Their shapes seem to slide under the surface as underwater islands. I'm ever more aware of the tentative land on which I stand, a white wound gouged out of England's underbelly, studded with flint and sutured by grass.

The chalk hurts my eyes. I lie back. The turf is springy and surprisingly comfortable, and worn out by my early rising and the afternoon sun, I fall asleep. I wake abruptly – with that disconcerting sense of not knowing where I am or how much time has passed – and realise I'm not in bed, but on the edge of a five-hundred-foot cliff. There are strident voices coming out of the air. '...*That's a risky business...*' says one. For a moment I assume they're talking about my perilous nap, but they soon drift by and I scramble to my feet.

As I walk on, the ground grows more fissured as the island narrows and the sea expands. The sheer unboundedness seems to invite me to throw myself off. 'It is as well, however, not to go too near the cliff-edge,' my 1950 guide cautions in solemn tones, 'as in this exposed corner the wind often comes with sudden gusts that might have awkward consequences.' I'm too close. I can hear my mother telling me to be careful, and see my father frowning when I persist in my daredevil ways. He feared for her heart. But then, so did I.

Here where the grass parts company from the white, chunks of cliff are preparing themselves for collapse, as if the whole thing might give way any minute. As the wind gusts enticingly around me, I wonder that the Poet Laureate and his photographer friend weren't caught up in its embrace; billowing cape, wide-awake hat, Indian shawl, red velvet dress, glass plates, manuscripts and notebooks, all sent flying over the edge as an airborne bundle to be dashed on the boulders below. As for me, there'd be no audience to my final piece of play-acting, only the birds, whose careless launchings encourage my fantasy of falling or flying, the same fear that feeds my apprehension in the sea. I trust to the water; they place their faith in the air.

As I stand there, I hear a sudden whoosh by my ear. A raven rises on the updraft, near enough for me to feel the wind from its wings. Riding on fingertip darkness, stark black against the bright white and wide-winged, it looks more like a bird of prey than a passerine, commanding the air around it. Landing deftly, it is joined by its partner, eyes glinting as they stalk the turf, a twitchy, mythological presence. Ravens, I have decided, are my new favourite animal.

If an animal's brain exceeds the size which the efficient running of its body would require, the excess is measured as an encephalisation quotient, or EQ. It is one way in which we can measure the capacity of the forebrains that govern sensation, memory and emotion. In the global clan

of corvids – crows, magpies, rooks, jays and jackdaws –
this index far exceeds that of all other birds. Above them
all is the raven, *Corvus corax*, the *über*-crow. It boasts
a brain-to-body size comparable only to primates and
toothed cetaceans such as dolphins and sperm whales,
even though a raven's brain weighs a fraction of an ounce
compared to the latter's eighteen pounds.

If we were once aquatic apes, owing our brains to our
seafood diet, then such is their intelligence that some
biologists go so far as to call corvids 'feathered apes'.
And as ever with science, one conclusion only invites
another in the endless cycle of what may or may not be
true. 'It may be impossible to prove in a literal or absolute sense that any one particular animal has or does not
have emotions, consciousness, or capacity for insight,'
as Bernd Heinrich, a biologist who has raised ravens by
hand in order to study them more closely, writes. 'These
subjective, individual, and hard-to-define qualities of

mind are found in separate independent evolutionary lines, with the highest end-points reached in some species of primates, cetaceans, and perhaps corvids and parrots.'

A carrion crow will place nuts on a pedestrian crossing for cars to crack, and wait for the red light to retrieve its meal. New Caledonian crows hook food out of holes with leaves and sticks, displaying the same insight and creativity that Aesop related in his fable of the crow and the pitcher, in which the thirsty bird learns to drop pebbles into the jar to raise the level enough to allow it to drink. Magpies recognise themselves in the mirror, suggesting a sense of individual identity. Jays remember the 'what-when-where' of past events. Rooks will support their fellow birds after a fight in a manner which in humans we would not hesitate to call sympathetic. These birds all demonstrate co-operative action which is clearly not instinctual. They are intensely social, bound by life-pairs and the kind of wider ties associated with cognitive animals – such as ourselves. Clever, knowing, tricksy, sometimes it seems the entire family are conspiring to hide their intelligence from us, for fear of what might happen if we found out what they really knew.

Despite their superior size, ravens are the most timid corvids, scared of a snapping twig, if not a passing cloud. They also deceive one another, pretending to bury food in one place, whilst surreptitiously taking it elsewhere. And unlike almost every other animal, they appear to possess the ability of 'gaze-following', taking note of where their peers are looking, and anticipating accordingly. Their 'observational memories' allow them to remember earlier crimes, assess the results of repeating them. In effect, they lie, like us; and like us, they also exhibit emotion – especially fear. They do well to remember who their enemies are, since they seem to have so few friends.

One raven will distract an eagle whilst the other steps in behind and steals its prey. Nesting ravens steal eggs from seabirds to feed their own young. Sometimes the relationship is mutually beneficial. A raven's call will summon wolves or foxes to a dead animal; the birds will then wait while the mammals pull apart the carcase, allowing them access to the meat. Such talents earned ravens the name wolf-birds; the Inuit would follow them too, led to their prey by the ravens' 'gong-like' calls.

Given such greedy behaviour, one might be forgiven for thinking that these birds must be the root of the word 'ravenous' – especially as they'll eat anything from offal to dogshit. In fact – although the word derives from the Old French, *ravine*, and before that, from the Latin *rapere*, to seize or snatch – the same root gives us rapacious and rape, and thus raping, which can also mean the act of tearing prey, and might have been made for these birds – which actually derive their name from the Norse, *hrafn*. Ravens can kill seal pups, reindeer and lambs, first pecking out their victims' eyes. They even stand accused of murdering one another. Such sombre crimes sustain their gothic air – yet in the past they were regarded not as harbingers of death and disaster but as companions or even begetters of creation itself; ceremonial birds, part of our rituals, as well as their own.

I must have seen my first raven on a childhood visit to the Tower of London, where they stalk the lawns with clipped wings, kept captive to warn of danger. The Roman founders of London believed the birds augured violent death and foul weather; and as they bore the characteristics of Saturn, they were a sign of that planet's ill-disposition: if the ravens left their nests, famine and calamity were sure to follow.

To blame human fates on a bird is as bad as Ahab investing a whale with evil. But like the whale, the raven has ever laboured under an elusive profile, one which shifts as fitfully as the animal itself. It presides over Christian legend: it was the raven, rather than the dove, which was the first to leave the Ark, searching for land and food, flying 'to and fro until the waters had dried up from the earth'; and far from signifying doom, ravens appear as servants to saints. Although the woebegone Job, a man never short of self-pity ('even young children despise me'), complained, 'Who provides for the raven its prey, when its young ones cry to God, and wander about for the lack of food?', his cheerier fellow prophet Elias was visited by ravens which brought him bread and meat in the wilderness, as ordered by Jehovah: 'You shall drink from the brook, and I have commanded the ravens to feed you there.' As a result, the bird recurs in Renaissance art, as well as in the words of the metaphysical poet Henry Vaughan:

> *Here* Jacob *dreames, and wrestles: there*
> Elias *by a Raven is fed*

One saint who found such succour was Paul, the first Christian hermit. As a fifteen-year-old boy, Paul fled from persecution into the Egyptian wastes, only to be followed by so many others that the wasteland became a virtual city of penitents. The Desert Fathers, as they became known, displayed extravagant acts of abstinence and denial. They faced the same devil with whom Christ had wrestled when he fasted for forty days among the wild beasts, cared for by angels. These first monks lived 'in a twilight between the real and the visionary', as Walter de la Mare wrote, starving themselves for weeks on end and even chaining themselves to the ground in an effort to outdo each other's piety.

As Helen Waddell records in her charming *Beasts and Saints*, a translation of their stories published, with elegant engravings by Robert Gibbings, in 1934, many Desert Fathers demonstrated a remarkable affinity with animals. St Mark the Wrestler cured a hyena whelp that had been born blind by spitting on his fingers and signing on its eyes. St Pachome walked unharmed among snakes and scorpions and summoned crocodiles to ferry him across the river 'as one calls a cab from a rank'. And St Simon of Stylites, who lived on a pillar for forty years, took a tree out of a dragon's eye, in gratitude for which the creature promptly turned Christian. Such acts evoked Eden, as Waddell wrote in an era which itself was rapidly darkening: 'In the first paradise that lies behind the memory of the world there was no cruelty...'

Rational thought might ascribe these scenes – for which there is a term, zooscopy, a form of mental delusion in which the sufferer sees imaginary animals – to isolation and malnutrition, but Paul thrived on just such a miracle. He settled cosily in a cave, wearing a garment of palm leaves, and until the age of forty-three lived off the fruits of the same tree. His diet doubled in variety when a raven came bearing half a loaf of bread – and repeated the feat every day for the next sixty years. Since ravens live to fifty or more, this is not entirely impossible, if one sets aside the other practicalities.

Still sustained by heavenly bread, the elderly Paul was visited by Anthony, another hermit who lived in the wilderness 'with no fear of the wild beasts which were therein', and who was himself ninety years old. That day the raven arrived with a whole loaf for the saints to share, an act of corvid catering depicted, sensationally, by Velázquez in a sublime painting that shows the bird flying out of the clouds and down a sheer cliff to the two aged eremites, clutching what looks suspiciously like a bagel in its beak. Evidently a diet of bread and dates was an aid to longevity: Paul lived to one hundred and thirteen, thanks to his reliable raven. At his death, around AD 345, he was buried by Anthony – who lived to one hundred and five – in a grave dug by a pair of friendly lions, as seen in the background of Velázquez's work.

The stories of the desert monks show how distanced we are today from animals. They evoke an age when beasts and birds meant more than just meat or servitude, since they represented the inexplicable wonder and fear of

the created world. It is why medieval bestiaries resemble typological tracts and religious analogies, with their islands that turn out to be whales, unicorns able to diagnose a maiden's virginity, and pelicans which pluck their own breasts to suckle their brood on their blood. The raven flits through such myths, shifting from creator and sustainer to destroyer and back again. Not long after Paul was interred in the desert, the mortal remains of the martyred St Vincent of Saragossa – who'd been roasted on a gridiron – were guarded from scavenging beasts by a raven. Later, St Benedict, the sixth-century founder of the famous order, was saved by a raven that snatched away a piece of poisoned bread which the saint was about to eat.

To Christians, the raven represented the immortal soul; some even saw in the bird's blackness a reflection of the brightness of the sun, just as the Roman followers of Mithras had seen the bird as a solar messenger. But more northerly beliefs began to darken its reputation. At the feet of the Norse god Odin sat two wolves, Gere the greedy and Freke the voracious, who fed him, while on his shoulders perched two ravens, Hugin, or Reflection, and Munin, Memory. These birds whispered in Odin's ears of what they had seen on their daily flying missions around the world, during which they occasionally stopped to drink the blood of wounded men. They were the enablers of Odin's omniscience and earned him the name of Rafnagud, the Raven God; his symbol was borne on the standard of his earthly armies, bent on their own predatory plunder. Unfurled on their banners, the raven was a harbinger of war. If it hung its wings, defeat loomed. If victory was imminent, it flew outstretched as though to warn its victims, 'This is what's coming for you,' shortly before turning them into carrion for the real corvids that would soon descend on the scene.

At sea, Norse sailors carried ravens as navigational aids as they sailed from one northern island to the other, as they would otherwise have become lost under the starless skies of high summer latitudes. Released, the birds would rise up looking for land. If they didn't find it, they'd return to the ship. The north was the home of such resonant creatures of the forest and the sea, their names eliding with internal rhythms – ravens, wolves, whales and bears – a zoomorphic cast endowed with all manner of semi-human characteristics. In his *History of the Northern Peoples*, Olaus Magnus maintained that there were 'extremely savage ravens, including white ones', that lived in the icy lands, capable of 'clawing out the eyes of babies as they lie squalling in their cradles'. He also claimed that the birds could make sixty-four different sounds, from *Cras, cras*, 'Tomorrow, tomorrow,' to *Erit, erit*, 'It shall be, it shall be,' although neither of these messages could be trusted. 'And so it is with all the other calls which ravens babble, cluck, bark, croak, gargle, and wheeze, telling of fierce and terrible storms, rains, and other disasters.' The raven was a liar and a trickster, the bird who cried wolf.

Such myths merged with the coming of Christianity. In early English churches, stone porches were carved with rows of raven beaks, reminders of Odin's servants as well as of the brutal behaviour of the tattooed Vikings, said to have flayed Christians alive and nailed their skins to church doors, as witnessed by fragments of human tissue found under nail heads driven into the oak. (My own ancestors may have been among those raiders: my mother's red hair inherited from her illegitimate grandmother, and my crooked little fingers, bent by Dupuytren's contracture, betray our Nordic roots.) Along with the wolf and the eagle, the raven was one of the Beasts of Battle, haunting emblems of death and destruction in Germanic, Norse and Old English verse. 'The Battle of Brunanburh' recorded corpses left 'for the dark/Black-coated raven, horny-beaked to enjoy'. In 'Judith', the raven is 'the slaughter-greedy bird' that rejoices at the coming feast. And in 'The Fight at Finnsburh', the raven circles, 'swarthy and sallow', over the fallen heroes: the original Anglo-Saxon expresses it better, in alliterative half-lines that emphasise the bird's eerie blackness: *Hræfn wandrade sweart and sealobrūn.*

St Oswald, the Dark Age king of Northumbria, was often portrayed with a pet raven which carried his ring to the Wessex princess he intended to marry. Later, the bird performed a posthumous service for its master, when in 642 Oswald was slain in battle and his body dismembered by Penda, the pagan ruler of rival Mercia. In order to propitiate Odin, Penda had the saint's head, hands and arms hung on stakes, but this macabre display had the opposite effect of spreading his opponent's cult throughout England. Oswald's raven flew off with one of his master's arms to a sacred tree – an echo of Yggdrasil, the heavenly ash which bound all time and space – where a holy well promptly sprang from the ground. One limb ended up in Ely, another in Peterborough, and in Durham, Oswald's head was placed in the coffin containing the incorrupt remains of a yet more revered saint, one with his own extraordinary relationships with animals.

There might not be much room in the modern world for monks and their miracles, but who could not love the story of St Cuthbert? His name alone sounds comforting, northern and true; he was also one of the most powerful men of his time. Born around 634, he grew up as an adopted child, and at one point seems to have served as a soldier. He entered holy orders in 651, partly as a result of the turbulent aftermath of Oswald's defeat and the death of the bishop, Áedán, whose soul Cuthbert had seen in a vision as it was carried up to heaven.

This fair-haired, athletic man, 'a conspicuous left-hander', grew to be as great a lover of birds and beasts as he was a devoted missionary to the north of England and Scotland. Stories of remarkable miracles would grow up in his wake, as related by his hagiographer, the Venerable Bede. Once, when stranded by a storm on a Pictish beach for three days, Cuthbert and his hungry brothers found three pieces of dolphin flesh laid out and ready to cook. (Conveniently, cetaceans could be eaten on fast days: the porpoise itself got its name from the French, *porc-poisson*, pork fish, and was ordained a 'royal fish' by the Normans to reserve it for the religious and the nobility.) On another occasion, while staying at the aptly named abbey of Coldingham on the Scottish coast, Cuthbert went out into the night to pray, wading naked into the water. Such chilly immersions were often undertaken by other northern monks to sustain their long vigils. One would often break the ice on the river to enter it. 'Is it not cold, brother Drycthelm?' his brethren called out, to which the phlegmatic monk replied, 'I have known it colder.' St David, too, was said to stand in the freezing sea to test his faith.

After spending the dark hours up to his neck in the water, towards dawn Cuthbert returned to the shore, to be followed by two otters which began to play around his feet, warming his toes with their breath and drying them with their fur. Having done so, they received the saint's blessing, then scampered back into the sea. (I wonder if the monks of Netley sought the spiritually-stiffening effects of Southampton Water? Perhaps if I were more saintly I might persuade a pair of marine mammals to perform a similar favour.) A twelfth-century illumination depicts the skinny-dipping Cuthbert cloaked by the waves, all the while spied upon by a curious fellow monk. Then, in a kind of time-lapse animation, the saint is seen seated on a rock, receiving his pedicure from the ghostly otters.

Northumbrian monasteries such as Coldingham, Whitby and Lindisfarne were built by the sea; it was their highway, their fastness, and their undoing, laying them open to Viking raids. Cuthbert sought somewhere less accessible, and found his desert island in an archipelago of thirty remote rocks, 'sieged on this side and that by the deep and infinite sea'. Nowadays the Farnes are famed for their grey seals, puffins, and nesting terns that threaten to peck interlopers' heads, but Inner Farne had been long haunted by dark-faced demons, clad in cowls and riding goats, 'their countenances most horrible'. The saint soon drove them off, much as his predecessor St Patrick ordered the serpents out of Ireland.

'Farne' means traveller or pilgrim, and in his quest for solitude, Cuthbert was not satisfied by the surrounding sea or the island's basalt buttresses. He built a circular

enclosure of boulders and turf whose floor he lowered 'by cutting away the living rock', leaving him with a view of only the sky, so as not to be distracted in his contemplations. Out in the unyielding grey of the North Sea, he became an anchorite on his island, if not anchored to it, like the chained saints of the desert, or the chained books in a library, both free and imprisoned in his cell.

Within Cuthbert's corral stood two structures: one an oratory, the other his house, roofed with rough beams and thatch. His settlement did not lack convenience: a third hut housed his toilet, handily flushed by the tides twice a day. And down in the island's harbour he built a hospitium where visiting brethren could lodge. At first Cuthbert, who would stay on Inner Farne for nine years, came out to greet his callers and would wash their feet, not unlike his otters; but latterly, he'd merely wave a blessing from his window.

In his lovely loneliness, Cuthbert sought only the company of non-human neighbours who, in return, performed services for him. Once, reading a psalter by the sea, the saint dropped his book into the water – I imagine its glittering illuminated and unchained pages fluttering as they tumbled into the murky depths, an expensive loss in an age when books were more precious than almost anything. At that moment, a seal dived down and returned with the book in its mouth. It too received a blessing for its efforts, although I suspect a little fresh fish would have been as welcome.

And in his self-sufficiency, Cuthbert discovered that sometimes the local wildlife had to be taught a lesson. When a flock of birds began to raid his newly planted field of barley, he reproved them, 'And why are you touching a crop you did not sow?' They were followed by a persistent pair of ravens, who stole from his roof to line their nest. The saint asked, patiently, that they should return what they had taken, only to be scoffed at by the birds. That roused him to anger. 'In the name of Jesus Christ,' he said, 'be off with you as quick as ye may, and never more presume to abide in the place which ye have spoiled.' They flew away, with a dismal look, ashamed of what they'd done.

Three days later, as Cuthbert was digging in his field, one of the pair returned, 'with his wings lamentably trailing and his head bowed to his feet, and his voice low and humble', as Bede relates. The bird begged forgiveness, and Cuthbert gave the pair permission to return. When they did, they brought with them 'a good sized hunk of hog's lard such as one greases axles with', which Cuthbert gave to his visiting brothers to waterproof their shoes. 'Let no one think it absurd to learn virtues from birds,' he declared, 'for as Solomon says, "Go to the ant, O sluggard, and consider her ways, and learn wisdom."' The reformed ravens remained for many years on the island, rebuilding their nest each year without recourse to Cuthbert's roofing materials, nor 'wrought annoyance upon any'.

Ravens were not the only feathered inhabitants of the Farnes that Cuthbert took into his care. Eiders, with their pistachio-green necks and wedge-shaped bills, their oddly reassuring call – *a-hoo, a-hoo* – and their solemn, sturdy presence, became particular favourites because of their seeming tameness: Cuthbert's successor on Inner Farne, Bartholomew, would allow the birds to lay their eggs beside the chapel's altar, and even under his bed.

In fact, eider females appear approachable only because they remain at the nest to protect their brood no matter how close a predator may get; it is one reason why raiders were able to steal their soft down. But Cuthbert's ascetic life had no need for feather quilts; instead, he instituted a law to protect the birds, the first such legislation in England.

At least, that's the story. It is not mentioned in either of the two earliest hagiographies of Cuthbert, although a fourteenth-century manuscript does refer to twelve pence paid to a Newcastle artist for painting *volucer S. Cuthberti* (medieval Latin for 'Cuthbert's bird') on Durham cathedral's reredos, perhaps prompted by Reginald's chronicle of the saint, in which he mentions 'certain creatures... named after the blessed Cuthbert himself'. Yet the legend suited what we knew, or wanted to know, of the man. It has persisted for a thousand years, earning the eider its endearment, Cuddy's duck, in honour of this northern St Francis.

After doing his duty as bishop of Lindisfarne, Cuthbert returned to Inner Farne, where he died in 687. Reluctant to leave the island in life, in death his remains, like Oswald's, were rendered refugee by invading Vikings, wandering from place to place in the care of his fellow monks. When Cuthbert's body finally came to rest at Durham in 1104, it was found to be miraculously intact, with his gold and garnet pectoral cross concealed in the folds of his cloak. It was even said that the saint's blood still ran fresh in his veins, and that he was gently breathing. His bird-entwined legend would remain potent, to be commemorated by another Newcastle artist, the Pre-Raphaelite William Bell Scott, whose mural at Wallington Hall depicted Cuthbert with terns hovering over his tonsured head and a loyal eider at his feet.

By that point, birds had assumed the shape of new familiars, from Coleridge's ominous albatross to Poe's gothic corvid, created by an author who took on the persona of The Raven, forever clad in black from frock coat to silk cravat and, according to Tennyson, 'the most original genius that America has produced'. Poe's 'ghastly, grim and ancient raven', always croaking 'Nevermore,' was

101

invested with the same sense of foreboding as Melville's White Whale. Both were inspired by *The Rime of the Ancient Mariner*, and both acted as fated antidotes to muscular Christianity or evolutionary rationality.

In an industrial era, the raven had come to embody a new, unnerving myth. In 1813 Caspar David Friedrich painted *The Hunter in the Forest*, a diminutive figure dwarfed by massive firs while a raven sits on a stump as an augury of defeat; Friedrich's *Raven Tree*, a tangle of branches and black wings against a lurid sky, conjured up a similarly gloomy scene. And in one of my favourite, if obscure, paintings, an 1868 watercolour by another Pre-Raphaelite and naturalist, Robert Bateman, the body of a dead knight lies in a shadowy forest glade in whose dark branches perch three ravens, 'as blacke as they might be', one saying to the others, 'Where shall we our breakefast take?' The story comes from an old English ballad, but the strangely submerged scenario could be set in some undefined future as much as in a romantic past.

Modern art was in no mood for cheery birds. Corvids appear as scurrying shapes in Van Gogh's *Wheatfield with Crows*, painted in the last weeks of his life in 1890, before madness and death overcame him, leaving his friend Gauguin to retreat to Tahiti, where in 1897 he painted his own dream vision of a nubile native girl overlooked by Poe's raven. Meanwhile, in Henri Rousseau's *War* of 1894, all the more macabre for its naïveté, a childlike figure rides a black horse over a field of dead bodies, among them one that resembles the artist himself. In the clutter of flesh, the crows peck and forage as if their bare and bloody beaks were descrying in entrails the Armageddon to come. Perhaps it was no coincidence that by 1918 there was only one raven left at the Tower of London, and its flock had to be replenished from the Dartmoor village of Sourton, over whose abandoned quarry the birds still swoop and soar, huge and black against the blue sky.

A thousand years after Cuthbert retreated to Inner Farne, Thomas Merton, monk and poet, sought to step outside the world to see it better, to find God within himself and within nature. Like Cuthbert, he had led a worldly life until then; a dissolute one, even, in the teeming streets and seedy dives of downtown Manhattan.

Born in France in 1915, son of a New Zealand painter and an American Quaker, Merton had an itinerant upbringing. His mother died when he was young, leaving his artist father to take him travelling: to the Cape Cod towns of Provincetown and Truro – 'a name as lonely as the edge of the sea' – where he first saw the ocean from windswept dunes; and to Rome and the south of France, where he felt his soul come alive.

After his father's early death from a brain tumour, Merton studied at Cambridge, where he drank and was said to have fathered a child. He left England in 1934, regarding it as a decadent place 'full of forebodings', 'a vast and complicated charade' whose people were 'morally dead'. His ship 'sailed quietly out of Southampton Water by night', leaving behind 'the silence before a storm...all shut up and muffled with layers of fog and darkness...waiting for the first growl of thunder as the Nazis began to warm up the motors of a hundred thousand planes'.

In New York, Merton enrolled at Columbia University, became a Communist under the assumed name of Frank Swift, and in between visiting nightclubs and playing 'hot' jazz records, studied William Blake (to whom he swore allegiance, perhaps on account of the fact that the poet saw angels in the trees of south London) and Gerard Manley Hopkins. Among his friends was the avant-garde artist Ad Reinhardt, who sought to create entirely black canvases – the ultimate artistic gesture of the age.

Veering from dissolution to devotion, Merton was twenty-three years old when he became a Catholic in 1938 – at the same time that my mother, then a teenager, was being received into the Church in Southampton. One morning, after staying up all night with friends drinking in a club, he suddenly realised, 'I am going to be a priest.' With war raging in Europe, Merton gave up his one-room flat in Greenwich Village for a cell in the Trappist monastery of Our Lady of Gethsemani deep in rural Kentucky, where he had watched a novice inducted into the order. 'The waters had closed over his head, and he was submerged in the community,' Merton wrote. 'He was lost. The world would hear of him no more. He had drowned to our society...' Having been rejected by the military draft, Merton burned all the manuscripts of his novels, gave away his possessions and money, and exchanged his 1940s suit for fifteenth-century underwear. He left modern America for a medieval enclave; the one-time Communist was given a new communal name, Frater Louis.

Merton's retreat was about being lost and finding a new home, a rebirth. Clad in brown cowl and white robe, rather like a bird, he declared, 'I desire to be lost to all created things, to die to them and to the knowledge of them.' The Trappists obeyed the rule of St Benedict, observing silence, speaking only when necessary. At first Merton was even forbidden from writing poetry, although by 1948 he had published his autobiography, *The Seven Storey Mountain*, which brought him to the world's attention, appearing in England under the title *Elected Silence*, edited by Evelyn Waugh and championed by Graham Greene.

Enclosed in his new order, Merton withdrew from the world and its problems so as to address them in the silence of his calling. Yet he could not resist speaking out. A contemporary of Auschwitz, Hiroshima and the Watts riots, he was the first Catholic cleric to protest publicly against the Vietnam War. He saw it as his duty to be, and in being, to reflect, in the manner of an artist. 'The monk is not defined by his task, his usefulness,' he wrote. 'In a certain sense he is supposed to be "useless" because his mission is not to *do* this or that job but to *be* a man of God. His business is life itself.'

Out of the contemporary cacophony – literally, shit sound – Merton listened to the voices around him. Hugin and Munin had gathered the news for Odin; this modern monk was God's radio receiver for what was going on in the world, and what was going wrong with it. When the marine biologist Rachel Carson published her book *Silent Spring* in 1962, exposing the terrible effect that pesticides were having on the natural environment, Merton wrote in support. He bore the same witness in his poetry and his photographs – 'The monk is a bird who flies very fast without knowing where he is going' – and increasingly looked east to Buddhism. In his book *Zen and the Birds of Appetite*, published in 1968, he drew comparisons between the Zen masters and the Desert Fathers, with their own relationships with animals. And reading *Moby-Dick*, he declared it had 'a great deal to do with the monastic life and perhaps a great deal more than the professedly spiritual books in the monastic library'.

The only known photograph of God

To Thomas, the raven – emblem of St Benedict – was the symbol of both salvation and mortality. Its black wings, as black as Reinhardt's canvases, might as well have been an augury of the nuclear explosions now taking place in another desert.

May my bones burn and ravens eat my flesh
If I forget thee, contemplation.

Merton's own end was abrupt, fiery, and shocking. He died while visiting Thailand in 1968, accidentally electrocuted by a faulty electric fan. He was fifty-three, the same age as Cuthbert when he died. Witnesses described his corpse in exact detail – the fearsome burns to his torso, contrasting with the placid expression on his face – as if he might be incorrupt too. His body was flown back to the United States on a plane carrying GIs killed in Vietnam. It seems a violent conclusion for a man of peace, for all that he'd appeared to predict his fate in the last line of his autobiography: 'That you may become the brother of God and learn to know the Christ of the burnt men.' There were even rumours that dark forces had conspired to do away with this troublesome priest.

By the mid-twentieth century the symbol of the raven had been perverted. In Norman Bates's motel, the shape of a stuffed raven looms over the shoulder of Miss Crane, an omen of her imminent murder at the hands of a psycho dressed as his own dead mother. But people did not need Hitchcock to scare them witless with scenes of psychotic corvids and gulls turning on the inhabitants of a seaside town. For generations they had suspected birds, and declined to have pictures of them in their houses. They were spooky, unpredictable creatures out of the shadowy past; lurking, dark-feathered contradictions; not symbols of beauty, but faintly repulsive and reptilian creatures, ready to turn back time and become the terrible lizards they once were.

Here, on the clear, pure heights of the island, is a landscape left to the birds. Up close and in reality, its ravens are huge, made bigger by their fluffed-up ruffs and feathered legs. I stalk one bird as it stabs at the earth, making myself look as ravenly as possible in my black anorak. If ravens are so clever, says a friend who lives on the island, how come they occupy such a limited niche, when you'd expect them to reign supreme? Having been forced from its inland home to more remote coasts, this alpha corvid is slowly recolonising the southern country it once knew well; and as with human beings, intelligence is not always a sign of success.

The raven pair are cold-shouldered by other birds, set apart. Their eyes give little away, saying nothing of whatever lies behind them. Maybe they're ready to lure me over the cliff so that they can feast on my bones below. Or perhaps they're taunting me for my ineptitude as they fly, banking and swooping and spinning and tumbling in the air, the sunlight shining through their primary feathers, turning them diaphanous and ghostly grey. They're joined by groups of rooks and jackdaws, summoned to a corvid convention. That summer I'll see their aeronautics mirrored by the distant loops of an air show. Perhaps we'd pay more attention to birds if they left contrails of their own.

As I watch, a new player makes its entrance: a peregrine falcon, riding on the updraught, streamlined and straight-winged – such a noble, Spitfire of a bird, so supercharged one might imagine it had a Rolls-Royce engine, even as its name evokes a chivalric past. In his book *The Peregrine*, a study in a reign of terror, J.A. Baker describes how these raptors 'perfect their killing power by endless practice, like knights or sportsmen' – a medieval analogy underlined by the name for a male falcon, a tiercel. Like the Farnes, 'peregrine' signifies pilgrim or traveller, since in the ancient art of hawking, newly-fledged birds were taken not from the nest, but while in flight from it. Its Latin name reflects this, as well as its sublime shape: *Falco peregrinus*, sickle-winged wanderer. Murderous and exquisite, it circles seemingly without effort, belying its facility as the fastest animal alive, able to fly at more than one hundred miles an hour, pursuing its prey with such velocity that a gull or jackdaw's head can be snapped off in the violence of its attack.

Baker – a librarian who lived in Essex in such obscurity that until recently no one knew the date of his death – followed the peregrine for ten years, a witness to its fenland fiefdom. He saw it living in a 'pouring-away world', negotiating the landscape in 'a succession of remembered symmetries'. During the writing of his book, published in 1967, Baker was diagnosed with a terminal illness. Yet his peregrine is no symbol of the uncanny; it is utterly of its world. It is a survivor, still recovering from the wartime culls when it was shot as a threat to carrier pigeons employed as radio silence was imposed on submarine-spotting planes, and from the second assault of pesticidal poisoning in the sixties, which nearly silenced it for good. Peregrines have nested on this island cliff for centuries, raising successive dynasties in eyries set into its vertical face. It is a place made for such a soaring performer, infinitesimally attuned to movement and space; 'the mastery of the thing!', as Hopkins wrote of another raptor. As the unholy pilgrim scans for prey, using eyes that are bigger than a human being's and five times more powerful, I might as well be watching a cheetah on an African plain as sitting on a cliff in southern England. To be alone with all this beauty seems somehow greedy.

This terrain may be managed by man, but it has been edited by the wind, funnelling up the Channel, clipping the gorse and biting it into bonsai hillocks. The bushes still give off their coconut scent, as if to lure Julia's husband from his palmy beaches. I'm almost bouncing along the springy turf, aware that any of it might collapse with my next step. Armpits and hollows, crevices and groins shifting like a restive sleeper under a downy duvet, all coursing through the ground, their cracks filled with lush plants that wouldn't survive out in the open. Close to the edge, the chalk has begun to break off in lumps like damp icing sugar. A slow-motion earthquake is sundering the island from itself. The whole thing is sagging and groaning under the weight of its natural history, ready to slip silently and solemnly away.

It all seems so gentle, this place, raised so far in the air. Heading west into the sun, I feel I could go on forever.

Abruptly, the rolling green gives way to an astonishing view, as though it had been thrown in my face: great chopped-up chunks of white rock launched out into the water, waves washing around their weedy feet.

I've seen the Needles ever since I can remember, but close to and yet still at a distance, they appear more strange than familiar, possibly because they're always changing. The three eroded stumps, like rotten molars, are the remains of the long-lost arches and towers that earned them their name. The missing fourth, a narrow, hundred-and-twenty-foot pinnacle, fell in 1764. If there'd been a locked-off time-lapse camera running on their geological demise, we'd realise how reduced these stacks are in splendour. Now they appear as mere props to the squat, red-and-white-striped lighthouse, with its double-occulting light. Like its counterparts around the coast, it has its signature pulse – *eclipse two seconds, light two seconds, eclipse two seconds, dark fourteen seconds* – as cryptic as a cetacean's clicks. It runs on autopilot, its tower flattened to accommodate visiting helicopters, watched from the neighbouring stacks by heraldic cormorants – once known inelegantly as eel-crows – and sleek black shags with aristocratic crests.

Behind lies other evidence of human occupation: man-made barnacles clinging to this land's end, relics of a past when West Wight was one big fortress, and when the entire south coast was studded with brick forts. Successive batteries were built into this slender white finger to defend all England from hostile bands, leaving it as riddled with concrete tunnels as the cliffs' burrows.

As I prop my bike by a chained gate, a pair of swallows swoop out of the darkness. Poking my head through the doorway, I hear insistent cheeping from a nest somewhere in the gloom.

At the foot of the cliffs are paddling guillemots, sharp-beaked northern versions of penguins. As with razorbills and fulmars, the island marks their easternmost breeding ground. They lie long and low like miniature black-and-white battleships, built for the open water where they spend their lives, only coming into shore to nest. As members of the auk family, they have an antediluvian quality, evocative of ancient engravings. As they flutter down from the cliffs, their wings look improbably small and stumpy in proportion to their barrel-shaped bodies. I can just about hear them, at this distance. The guillemot 'utters queer and eerie noises', says my *Pocket Book of Birds*, published Spring 1936, 'reminiscent of the moaning of a person in pain'. Like so many British seabirds, they have greatly reduced in numbers in recent years. Puffins too once made their nests here; they have long since vanished, although later that summer I see a sole specimen far out at sea.

For centuries such birds were taken in their thousands for their meat, their oil, or their feathers. So great was the demand that the wings were torn off wounded birds which had been shot; like definned sharks, they were thrown back into the sea to die, in order that fashionable ladies could walk about with kittiwakes on their heads. They may have lacked a Cuthbert to defend them, but in my imagination I see spectral flocks pursuing those

Bond Street dames, demanding the return of their rightful property.

In the summer of 1936, as my *Pocket Book of Birds* appeared, T.H. White was busy training a goshawk. His friends found this ridiculous, and told him so. ' "Why on earth do you waste your talents feeding wild birds with dead rabbits?" Was this a man's work today? ... "To arms!" they cried, "Down with the Fascists, and long live the People!" '

White's account of his relationship with Gos – which would not appear in print for fifteen years, and which went under the working title 'A Sort of Mania' – is a ferocious but oddly opaque account, quite as obsessive as J.A. Baker's. It is bound up in medieval references and veiled allusions to the demons that drove this handsome man to become a poet and writer – as well as an artist, a hunter, an aviator and, perhaps most unlikely of all – and that almost by accident – a kind of pacifist.

White's journals of that testing time – which he called his day-books – are filled with loving drawings and tiny photographs of his goshawk. He even Sellotaped the bird's moulted feathers to the pages, just as he fixed a salmon scale to an earlier journal about fishing in the Scottish Highlands (and proposed that every reader of the published work should receive a similar scale with each copy of the book). The fact that White lost Gos halfway through its training adds pathos to these relics. The bird flew off with its leather jesses around its feet, never to return. Unsentimentally, White concluded that the hawk on which he had lavished so much attention – to the point of staying awake for days and nights, 'walking' the bird so that in its own fatigue it would bend to its master's will – had died with its jesses caught in some distant tree, where, 'hanging upside down by the mildewed leathers, his bundle of green bones and ruined feathers may still be swinging in the winter wind'.

Terence Hanbury White was born in Bombay in 1906, a product of the Raj as much as Julia Margaret Cameron, and as adrift from birth as Thomas Merton. Sent back to England by his uncaring parents for his education, he too gravitated to Cambridge, where, among other talents, he determined to learn medieval Latin shorthand so that he could translate the bestiaries which would come to influence his own work. Having published his poetry, he considered writing a biography of Gerard Manley Hopkins, but left his post as a schoolmaster at Stowe – where he let loose grass snakes in his sitting room and sunbathed naked on the lawns – to retreat to a five-shilling-a-week gamekeeper's cottage on the Stowe estate.

Here he lived in rural solitude, drawing his water from a well and using an earth closet – although he also spent a hundred pounds on carpets and curtains, mirrors and an ornate antique bed, and stocked his larder with tinned food and bottles of fine Madeira. In 1936 he published an account of his time there, archly entitled *England Have My Bones*, in which he sought to define himself, and the morally empty country Thomas Merton had left two years before. 'Nowadays we don't know where we live, or who we are,' he wrote. 'This is why, in a shifting world, I want to know where I am.'

His book was assembled from his various hunting, fishing and flying diaries, with one eye on his ancestor, Gilbert White of Selborne, and the other on Richard Jefferies, the Victorian naturalist who recorded the vanishing world of southern England and who wrote 'To me everything is supernatural' as he lay on his back on the downs 'so as to feel the embrace of earth', imagining the sky was the sea. White rhapsodised, in similar tones, about the country that he knew, caught in a few years of peace, a kind of insular refuge in time. '*20.iii.xxiv.* So is the whole British Island an anchorage, if you avoid the towns. So are birds and beasts and the sporting seasons... All the *things* which will outlast London are important to philosophic man.'

White may have been thinking of Jefferies' futuristic novel *After London, Or Wild England*, published in 1885, in which the writer imagines how the country would look if its capital ceased to exist and the land returned to its natural state. White could not know, though he may have suspected, that the coming years would see that city threatened with just that sense of oblivion – or that the countryside too would change as radically. And as William Cobbett, on his rural rides around Hampshire, had delivered a similar polemic against the iniquitous effects of the Industrial Revolution in the early nineteenth century, so White's was a last glimpse of Britain before the mechanisation to come, when its fields would be turned into chemically treated food factories. He wrote of its people and its animals, and of the carrion crow which he believed to live 'as long as a man, is extremely destructive of game, and is hunted for that reason with much enthusiasm by gamekeepers, as well as by emotional people who object to the creature for pecking out the eyes of dying animals'.

In fact, crows live for ten years, often less. But to White, animals were barometers of an even greater threat, unable as they are to defend themselves. He wrote in a weird interregnum, an era which flirted with any creed or philosophy or politics, no matter how extreme – indeed, made more so by the bookends of global conflict just past and soon to come. White made up his own myths, his own heroic narrative. He hunted and fished and learned to fly. 'Because I am afraid of things, of being hurt and death, I have to attempt them,' he wrote. 'This journal is about fear.'

It is not surprising that his book ends violently, as White leaves friends late at night after drinking, careers off the road and crashes into a ditch. His head strikes the dashboard of his car, and his nose and throat fill with blood. As he gets out, into the still-shining beams of the headlights, he realises he has lost the vision in one eye. Happily, he regained it, and resumed his helter-skelter life.

White could not stay still. He despised people who 'don't do enough things with their bodies', and was proud of the fact that he had not slept in London for five years. He railed against the countryside's despoliation – 'perhaps one day the New Forest will be the name of a tube station' – and said that the best thing for Britain would be a new war to wipe out two-thirds of its population. Yet White would spend the coming years in exile in Ireland, a kind of conscientious objector by default, having been advised against enlisting by his friend and fellow hunter Siegfried Sassoon, who told him three days before the Munich Agreement in 1938, 'The only way to be helpful in this emergency is to remain as calm as a wick and to keep still' – much the same advice White had given himself when dealing with his beloved but infuriating hawks.

You can see why White and the author of *Memoirs of a Fox-Hunting Man* might bond over dead animals. But White was much less enthusiastic about killing animals out of print. He found it difficult to sacrifice live birds to his hawk, regretted the death of a mouse, and above all doted on his red setter, Brownie, 'my Pocahontas, my non pareil'. She sat on his lap when he wasn't working, and groaned at the typewriter that took her place when he was. While her master hunted geese in the freezing winter dawn, she wore a flannel coat, her hindquarters in his bag, his woollen mittens on her front paws. In the rain, she sported a waterproof coat with spatterdashes to stop her feathery legs getting mud all over the house (a canine costume which, when he moved to Ireland, was the source of suspicions that White was a spy, carrying secret maps stashed in his dog's coat). And had anyone ever harmed her, they would have faced swift retribution. When out with a hunting party, White made it clear that if anyone shot his dog by mistake, it would be the last one that they made. 'I shoot him. Shot one. Shot two. Like that – no hesitation.'

White might have gone on in this vein, ever more mis-anthropic and reclusive, fading into a grumpy literary footnote. But that same year, his life was overturned by sheer luck, as he called it, likening his success to 'winning the pools', going almost overnight from living on credit to being a rich man. In August 1938 the American Book Club selected his novel *The Sword in the Stone* – Arthurian romance re-imagined for a troubled age, just as Tennyson's and Julia Margaret Cameron's works were for theirs. Since his days at Cambridge, White too had been fascinated by Malory's *Morte d'Arthur* – a story with a new relevance, despite being sourced in medieval and Anglo-Saxon legend. White's work would extend into five separate volumes. It was as much an act of medium-ship as of writing. 'I am trying to write of *an imaginary world which was imagined in the 15th century*,' he told his friend Sir Sydney Cockerell. 'I am looking *through* 1939 *at* 1489 itself looking *backwards*.' He accompanied this with a sketch of himself looking through a telescope, as if through time. Perhaps he saw the black knights as storm-troopers, Camelot as a bunker, and the merlins as fighter planes, while the white cliffs of Albion, Dover and West Wight became England's first and last defence.

White's book was written as war became inevitable. He was both caught up in the spirit of the times, and set outside it, too. Thirty years later it would catch my own imagination when I took it out of the little public library across the road from my school, where our father took us on Thursday evenings. I identified with its boy hero, Wart, the once and future king, a prince-to-be who becomes a hawk himself, soaring over an invented, idyllic Middle England; I saw myself pulling the sword from the stone, as page to a knight, a boy-soldier.

White never lost his own boyish enthusiasm, his sense of self-invention. He was, in the words of his bio-grapher Sylvia Townsend Warner, 'more remarkable than anything he wrote'. He designed his own logo, a flying hawk, and in his *Who's Who* entry would list, 'Recreation: Animals', later extending it to include painting and hawk-ing. For him birds were not pets or prey, nor even under his dominion. He did not tame his hawks: he entered into an uneasy truce with them. So too was his relationship with the world. At that moment, as Thomas Merton was contemplating monastic life in America and would soon receive his summons to the draft, conscription became a serious possibility in Britain, 'and everybody lives from one speech of Hitler's to the next', White wrote in his diary for 26 April 1939. 'My nature is not monastic; it may be noncooperative, but it is free. It is a raptorial nature. Hawks neither band themselves together in war, nor yet retire from the world of air.'

War for White came as a suffocation, symbolised by the day a farmer friend came to help fit him with a gas mask. Warner sees her subject trembling like an animal as the rubber seal is put over his face – before tearing it off and running into the woods. White declared he would neither fight nor run. 'Anybody can throw bombs,' he said; he had novels to write; it was his destiny. He even claimed that the overarching theme of *Le Morte d'Arthur* was to find an antidote to war. In Ireland – having left England before the issuing of identity cards or ration books and being beyond conscription age anyway – he gave up his identity just as Thomas Merton did; only he used the royalties from his writing to rent a large house and live in a medieval manner. Not for nothing did he tell Cockerell – who was the well-connected director of the Fitzwilliam Museum in Cambridge, and who had begun his own career by sending seashells to John Ruskin – that he had to try on a suit of armour in order to enter Arthur's world: 'I want to know how the stuff *works*.'

It may have been easier to re-imagine that world in the Celtic twilight of Ireland. White retreated into his island fastness. From there he corresponded with figures as disparate as Noël Coward, as to the suitability for the stage of his Arthurian play *The Candle in the Wind*, and Julian Huxley, on whether animals had a 'mind'. Reviewing Huxley's *Kingdom of Beasts*, in which the renowned zoologist claimed that man was the mammal

'most successful at living', White countered that 'Man...
has only his own word to go on, and to be most successful at what you happen to be best at doing does not constitute an absolute superiority over the rest of the animal
kingdom.' He would rather have lived with animals.
'How restful it would be if there were no humans in the
world at all,' he observed. 'If only there was a religious
order which not only took a vow of perpetual silence but
also decided *to go to bed* for ever, how gladly I would
join it.'

White's world was closed down by war. He was
as captive on his island as any internee on the Isle of
Man. He could not leave Ireland, nor enter England.
Communication with friends such as David Garnett (who
himself had been a conscientious objector during the first
war, when he was lover to Duncan Grant) and Sydney
Cockerell was reduced to fitful correspondence. In an age
already reliant on telephones and telegrams, radio and
radar, White's words were restricted to letters and diaries,
in the past of the old England he imagined. He thought
seriously about becoming a Catholic, went to Mass
every Sunday, observed abstinence and even considered
becoming a priest as an alternative to being a combatant.

But as he wrote his epic story of battles and ancient
heroes, White realised that by charting Arthur's story he
had sealed his own fate; that he had no choice but to enter
the war himself. To remain outside it would be to betray
his creation, let alone the country which he determined
should have his bones. He saw direct parallels with the
animals he loved. In nature, he claimed, only ants and

bees fought wars. Animals did not own property, nor conduct industry. They avoided rather than courted aggression. 'Now what can we learn about abolition of war from animals?' He hunted, but he only killed the things he loved very much, which is why he was not keen on killing human beings.

Deep grief brought drastic thoughts, and what he learned from the death of his beloved Brownie 'maimed his heart'. His red setter had become as eccentric as her owner-lover. She would adopt chicks and baby rabbits and bring them to bed with her; and since she and her master shared the bed, he would end up being bitten by rabbits. Brownie also kept a collection of stones under the kitchen table, to which she added regularly. And when she died, on the one day that White was away on business in Dublin, her master was so distraught that he stayed up for two nights with her body beside him, only then burying her in the garden, having clipped a lock of her golden-red tresses to tape into his journal.

For a week afterwards he visited her each night to say, 'Good girl: sleepy girl: go to sleep, Brownie', familiar phrases to reassure her, since he half-believed that her consciousness might persist. It was almost madness, he admitted, 'but that was the kind of chance I had to provide for'. 'She was the central fact of my life,' he averred; the only being he dared to love. He blamed himself for his absence when she died, for having failed her; for having killed the thing he loved more than anything.

It was too late, too, to enter the fight. By now the war had passed, and with it White retreated yet further. He moved to the Channel island of Alderney, partly to escape British taxes, partly because it was the only island that would accept his new dog, Killie. There he discovered the sea: 'You cease to be your own master, as you become a waiter-on upon the moon and sun, whose tides take no account of mealtimes or bedtime or times to get up.' And there, in 1959, he was interviewed for BBC television by a precise young presenter named Robert Robinson.

Forever smoking a pipe, White was now white-haired with a beard and a sailor's tan, looking older than his years: he was only fifty-three. Living on an island had served to accentuate his peculiarities. He painted the interior of his house red, and often wore a cassock of scarlet towelling. He resembled a more epicene, English Hemingway: the film begins with a still of White raising a shotgun over his head; his eyes twinkle with mischief and teasing. He tells Robinson that it is part of his physical training as a writer every day to 'swim underwater looking at fish'. He had recently taken advantage of a visit by an Admiralty diving vessel to don an old-fashioned helmet and rubberised suit, which reminded him of what it was like to wear a suit of armour.

'I met another diver at the bottom and we leaned against each other like amorous manatees... It was enchanting to be mothered by these tough, tender, bronzed young men,' he wrote, while the sailors dressed him in his suit like squires attending their knight. In the garden of his nineteenth-century villa he'd dug a great pit and built a swimming pool which he referred to as a gladiatorial arena and in which, in the film, we see two young boys diving. Close by he was constructing a temple to a Roman emperor – 'I think Hadrian was a very fine fellow' – and in between, besides recording the commentary for a film about puffins he'd made, took the time to paint vaguely surrealist canvases of female nudes and disembodied eyes. He seemed to be happy in his loneliness. Or at least, this is what the camera saw.

White was a product of a particular breed, not unlike another naturalist, Henry Williamson, author of *Tarka the Otter*, who lived in North Devon where he slept in the open and swam naked and believed in an English race-memory of divine 'ancient sunlight', as well as the power of fascism (in which he sought to enlist T.E. Lawrence, who carried Malory's *Morte d'Arthur* on his military campaigns, and who was also a good friend of David Garnett's). White, however, declined to declare for any political party. 'I don't think about myself very much,' he told the camera, constantly jabbing at his face with the stem of his pipe as if in self-defence. 'I don't know what I am.'

Admitting that he is a 'childish man', White complains to Robinson – his captive audience – of being typecast as a 'whimsical' writer. 'When I was young, you had to be grown up,' he says. 'It was fighting talk to be an escapist.' He was, he said, 'a middle-class, Edwardian Englishman'. He had no need to kow-tow, as he put it, to the modern world, which he accused of forever banging on drums to drown out its fear of the atomic bomb. He was insular and insulated by the success of *The Once and Future King*, then in the process of being turned into a musical, *Camelot*, the words of which I'd memorise from an album bought for my birthday. In White's case, he had Julie Andrews as a houseguest to sing its songs.

To the BBC, White presented a contrasting but happy picture of himself, an English writer in island exile, outside the reach of the Inland Revenue. Yet his life on Alderney had become complicated by something beyond his control. He had fallen in love with a young boy whose family had been visiting him. White did all in his power to entertain the boy – known only as 'Zed' – on his island, conjuring up its wonders like Prospero. Zed remained unnamed in Sylvia Townsend Warner's remarkably frank biography of White, published in 1967. But it is tantalising to wonder if he is the same boy we see in that black-and-white film, under the summer sun; and to wonder if White saw him as his young self, the squire to his knight, the Wart to his Merlin.

The relationship, which remained entirely 'respectable' on White's part, ended unhappily. Realising its impossibility, he cut off all contact with Zed, and was left with a wounded heart. 'All I can do is behave like a gentleman,' he wrote at the end of the summer. 'It has been my hideous fate to be born with an infinite capacity for love and joy with no hope of using them.' Despite the money and magic his books had brought him, they could do nothing to protect him from himself and his human emotions. His vivid imagination and innate sense of isolation – the essential unreality of his life which had produced them – only made matters worse. White's retreat was nature. Animals do not answer back.

Five years after this interview, White died on his way home from a lecture tour of America; he wrote his last letter as the ship passed the Azores. He was found in his cabin a few days later, having suffered a heart attack. 'I expect to make rather a good death,' he had told David Garnett. 'The essence of death is loneliness, and I have plenty of practice at this.' On 20 January 1964 he was buried in Athens. England would never have his bones. Nor would he live on in Albion, although, in some versions of the Arthurian legend, when the mythical king died, he turned into a raven, regarded in the West Country as a royal bird.

High on the downs, I watch the water in which I'd swum. The sea writes its own story, forever coming in and going out, entering and exiting, remorseless in its attack, knowing we are ultimately helpless in the face of its power; the same helplessness to which we yield, like grateful lovers, softly eroded and worn down in its crumbling embrace.

But this island is not a retreat for everyone. On another ferry journey, later in the year, I return to a high-walled prison complex, where men are held whose crimes are deemed so awful that their fellow convicts put broken glass in their food. They are confined in cells no bigger than a saint's, while out in the courtyard, its garden beds bright with marigolds, stands a wire aviary filled with birds. Inside the prison chapel, its concrete roof slowly leaking rain into a bucket, I talk to the class about writing, and they perform their exercises like grown children.

They write of birds and animals and the sea and the places they can sense, over the walls and in their past.

As I ride back over the green ridge of rising chalk, the shadow of Tennyson's cross falling behind me. In the distance, the descending sun sets off a semaphore flash from the windscreen of a far-off fishing boat. An hour later, the ferry delivers me back to the mainland. Terns wheel in our wake, continually plunging into the white water, and the light is fading fast as we pull into the dock.

3
The inland sea

Merely to be alive, indeed, is adventure
enough in a world like this, so erratic
and disjointed; so lovely and so odd and
mysterious and profound. It is, at any rate,
a pity to remain in it half dead.

WALTER DE LA MARE, *Desert Islands,* 1930

The suburbs slip by in a succession of deserted stations, trailing disembodied voices from unheard Tannoys. Silver birch march beside the tracks, shiny as wrinkled tin foil. Plastic bags blow in sycamore branches. The train rolls on. A dead badger lies slumped between the rails; a fox ambles out of the undergrowth. Despite the splintered chairs and decaying mattresses dumped over walls and wire fences, it would take only a few weeks for nature to overcome it all, after we've gone.

Ever since I lived there, leaving London felt like an escape. It frightened and excited me with its streets and alleys down which I might disappear, the soulless suburbs surrounding its dark heart, their entangling avenues like some voracious octopus. I used to think of how I'd get home if war was declared. How, once the sirens had sounded or the newsflashes came over the radio, I'd have to walk through the endless outskirts, along a motorway or down the back roads, trying to find my way south.

I was defeated and enthralled by the scale and sprawl of the capital which spread like a stain from its river. I descended into gloomy basements made glamorous by punk and all that came after, caverns in the dreariness, sparking into energy and abandon. I learned to drink and take drugs and dress for each night as if it was the first and last. I wanted to relive the posters on my wall, to escape my origins. And yet I always wanted to go home.

London, skewered on its own waterway, represented containment. But to set yourself free you had only to get on a bus, one of the routes that might have been running since the city began, carrying ghostly passengers through the ever-changing, never-changing streets, immemorial generations told off in bus stops. From the top seat of a trembling double-decker the familiar panorama unfolds: Old Street to City Road, the Bank to St Paul's, Blackfriars to the Thames. New shops and new skyscrapers rise, high-water marks of the city's fortunes. One morning I was all but thrown out of bed in my flat by an IRA bomb; now the steel and glass has grown even higher over the ever-erased, ever-rebuilt city.

From this jerky eyrie you see everyone undone. The bus creeps through the streets, its obstinate slowness a rebuff to the manic movement around it; the sclerotic traffic in its blue haze; the people on their way to work, briefcases and coffee cups in hand, sneaking surreptitious cigarettes in illicit clouds of smoke.

Along this ramshackle route lies my history, too: Bunhill Fields, a green refuge on a summer's afternoon, though under its turf lie the bones of William Blake and thousands of victims of the plague; Barts Hospital, in whose Victorian interior I was operated on for a strange white patch that had begun to spread along my spine, elaborately diagnosed as *Lymphangioma circumscriptum* as if it were written on my back and which, once excised, left a scar running like a thread knotted through my vertebrae; the newspaper offices in Fleet Street where I worked shifts wearing a cheap suit in the last empire of typewriters and alcohol; and Holborn Hill, where, in the opening of *Bleak House*, Dickens fantasised about a muddy city 'as if the waters had but newly retired from the face of the earth', with 'a Megalosaurus, forty feet long or so, waddling like an elephantine lizard', fresh from the Thames's ooze. All along the way are monuments to forgotten men, their dreams disguised by voluminous beards, and muscle-bound tritons, their scaly thighs rising out of the water that runs beneath the streets. It is strange to live in a place where rivers are forced to flow underground: Holborn itself means hollow bourne or brook, one of the streams that fed into the river Fleet.

An artist, who also happens to be a diver, told me how, some years ago, he'd lived in a dank basement in Pimlico. The atmosphere of the place was almost oppressively cold and damp; he was reduced to wearing blocks of polystyrene on his boots to keep his feet warm, with a similar layer under his bed. One day he looked outside and saw that a large portion of the brick pavement had fallen away, like a piece of jigsaw puzzle, to reveal free-flowing water; not in a drain or a conduit, but with gravel like a proper river bed. It was as if someone had punched a hole in the city's carapace.

Like so many cities – Venice or Hong Kong, New York or Amsterdam, St Petersburg or Mexico – or sacred sites such as Winchester, whose cathedral foundations were saved by the Edwardian diver William Walker, or Ely, the ship of the fens, sailing on the marshes through which slither its eponymous eels – London is an illusion; it only floats on sand and clay. Nowhere is beyond the tidal reach; the river brings the sea into the city. Sometimes it might even be invited in, as it was at Sadler's Wells, whose Aquatic Theatre, which opened in 1804, was fed by the New River – itself a diverted waterway – and boasted a sunken tank measuring ninety feet long, twenty-four feet wide, and three feet deep.

Here one could witness the Siege of Gibraltar and Neptune's chariot drawn by seahorses, along with other 'perilous and appalling incidents': a woman falling from the rocks to be rescued by her lover; sailors leaping from a vessel on fire; a child thrown in by its nurse, who had been paid to drown it, only for the infant to be rescued by a Newfoundland dog. So affecting were these scenes that at the end of a performance, members of the audience would jump into the water to assure themselves it was real. A theatre in which one could swim was remarkable even for the greatest city in the world, to be rivalled only by the dolphinarium which opened in Oxford Street in 1971, a murky green cellar-pool twelve feet deep, complete with swimsuit-clad 'aquamaids', a sea lion, a penguin and a trio of dolphins named Sparky, Bonny and Brandy who were prevailed upon to wear plastic hats and perform the usual tricks.

The modern capital barely acknowledges the river which was the reason for its being. The Thames was not embanked until the mid-nineteenth century; you could not walk or ride along its banks, yet you could descend to the water via a series of steps, marked on old maps – Temple Stairs, Essex Stairs, Arundel Stairs, Surrey Stairs, Salisbury Stairs, Whitehall Stairs – while street names such as The Strand commemorated what was once a beach. The river was much more present then; it seeped into the city, washed the feet of its buildings. Now the embankment oversees it with a wonderful disconnection – all those cars racing by, the trains rolling over its bridges, the commuters pacing to work; all turning their backs on the strong brown god as it surges through the city. It may be the colour of mud, swirling with silt and underlain with every kind of refuse, but I'm tempted to take to it like Benjamin Franklin, who, while working as a printer's apprentice near Lincoln's Inn Fields in 1725, would frequently plunge in.

Franklin was obsessed with swimming. 'I had from a Child been ever delighted with this Exercise,' he wrote in his *Autobiography*, 'had studied and practis'd all Thevenet's Motions & Positions, added some of my own, aiming at the graceful & easy as well as the Useful.' Even on his way to England, watching porpoises, grampus and dolphins as his ship passed the Isle of Wight, Franklin jumped in and swam around the boat as if to join them. In London, he taught a fellow apprentice – 'an ingenious young man, one Wygate'– to swim too. On a boat trip to Chelsea, Franklin decided to demonstrate his skills to the company on board. He stripped off and leapt into the river, then swam back to Blackfriars, 'performing on the Way many Feats of Activity, both upon & under Water, that surpriz'd & pleas'd those to whom they were Novelties'. He even considered starting a swimming school in the city, before fate took him to other things.

Sometimes, in the great stretches of my unemployed life, I'd get off the bus and walk the narrow alleys of the City, through the old inns of court and their silent celebration of privilege, being, on all accounts, an outsider. The lanes led to Lincoln's Inn Fields, its shrubbery home then to the down-and-outs and dispossessed of the 1980s who had turned it into an encampment, stretching their plastic sheets underneath the massive plane trees. On one side of the square sits Sir John Soane's house, its antique plaster casts and bizarre lighting effects set in a sepulchral interior; on the other rises the unremarkably grand façade of the Royal College of Surgeons, with its own cabinet of curiosities.

After climbing a wide staircase lined with portraits of past presidents, the visitor is greeted by a quartet of worn wooden boards on the wall. They appear innocuous enough, until you learn that they are overlaid with the nerves and blood vessels of human beings. This macabre furniture suite was the property of the diarist John Evelyn, who acquired it in Padua in 1643. Its planks are the essence of bodies; everything else – flesh and fat, bone and offal – has been stripped away to leave these upended tabletops, their grain coursed by tributaries like congealed river systems. They act as a ghoulish advertisement for what lies beyond: an array of glass cases on whose shelves sit innumerable jars and bottles and boxes, containing every imaginable body part, both human and animal. It's as though the world had been turned inside out and into its constituent parts by their collector, in whose honour the museum is named.

John Hunter, born in Scotland in 1728, was a surgeon and anatomist who trained at Barts. After serving in the army, he set up his practice in London; his brother William was obstetrician to Queen Charlotte, and delivered the future George IV. As a fashionable doctor, John Hunter was in demand. His patients included an older and wiser Benjamin Franklin, now suffering with a bladder stone, and the future Lord Byron, whose birth he attended and for whose club foot he prescribed a corrective boot, advice which went unheeded by the poet's mother. For both of these patients, as enthusiastic swimmers, Hunter's procedure for dealing with the nearly drowned might have come in useful: he recommended electrical stimulation to restart stopped hearts.

Hunter's curiosity knew no bounds. His motto was to experiment, rather than merely ask questions. He was the first to scientifically describe teeth, and transplanted them, still living, from one patient to another. He even experimented on his own body, dipping his lancet in the lesion of a prostitute, then incising his glans and prepuce, aiming to inoculate himself with gonorrhoea, only to contract syphilis too. Less dramatically, he also commissioned his pupil Edward Jenner to take the temperature of a hedgehog.

Hunter was a great teacher, but his blunt-speaking manner did not endear, nor did his apparent thirst for blood: Blake portrayed him as 'Jack Tearguts' in his satirical burlesque *An Island in the Moon*. In London, Hunter lived in a grand house on Leicester Square, where guests were greeted in his dining room by the gilt-framed specimen of an erect penis; but his country residence – to which he would drive in a cart pulled by a buffalo – was in Earl's Court, described by *The Gentleman's Magazine* as 'about a mile beyond Brompton, in the midst of fields', a rural scene – inhabited by 'animals of the strangest selection in nature'. Here Hunter was surrounded by a living museum, one giant experiment-in-progress,

> *On gizzards of gulls, hawks and owls,*
> *The heat of lizards, spurs of fowls,*
> *Bones of pigs, air-sacs of eagles,*
> *Moaning dingos, barking beagles;*
> *Sleek opossums, prickly hedgehogs,*
> *Buffaloes, dormice, wolves and dogs*

Leopards and jackals lurked in dens, zebra and ostrich roamed the lawns. There were eagles chained to rocks, buffaloes in the stables, and giraffes nibbling the trees. There was also a boiler for rendering down carcases, both animal and human, and an underground cell which I hope did not contain live specimens – although, if the engravings of the menagerie at Exeter Exchange are to be believed, I suspect it probably did.

Menageries had been part of London life since the thirteenth century, when King John kept his collection at the Tower, including Barbary lions whose remains have since been identified as an extinct species, as well as elephants, leopards and a polar bear which was allowed to fish in the Thames for its food while tethered by a long chain. In the eighteenth century, exotic animals excited a city alert to every new sensation; they were a kind of theatre. At the 'Change, as it was known, bonneted ladies and high-booted dandies were entertained by lions, tigers, monkeys, crocodiles, sea lions, rhinoceros and even an Indian elephant named Chunee, all displayed in cages barely bigger than their captives' bodies and housed on the upper floor of a building on The Strand, a veritable department store of beasts. Horses passing on the road below would rear up at the roars of the lions above, and a zebra was once ridden from there to Pimlico.

Visitors included Jane Austen and Lord Byron. The latter recorded: 'The elephant took and gave me my money again – took off my hat – opened a door – *trunked* a whip – behaved so well, that I wish he were my butler.' In 1826, after accidentally crushing his German keeper

when he turned in his cage, Chunee, suffering a septic tusk, tried to break open his iron bars and shook the walls so that the owners believed the entire menagerie might be set loose on London's streets, 'there being several lions, tigers and other ferocious beasts confined in the same apartment, all of which he might easily have liberated'. Soldiers from nearby Somerset House were summoned, and, as its keeper ordered his obedient charge to kneel, fired into the elephant's cage, while the other animals growled. 'The animal, finding himself wounded, uttered a loud and piercing groan,' and tried to free itself with its trunk, then hid at the back of its cage, only to be stabbed by long spears.

After 152 musket shots – and with crowds assembling outside and people offering to pay to see the beast die – Chunee had to be finished off by the keeper with a harpoon. His demise was as chaotic and agonised as that of any hunted whale; his remains were equally coveted. His carcase was dissected by students from the Royal College of Surgeons and his skeleton later put on display there, while his hide was publicly auctioned and his meat sold, along with recipes for 'elephant stew'. Letters of protest to *The Times* prompted the establishment of the Zoological Society of London that year, as a more enlightened way of keeping animals. 'To place an elephant, or any beast, without a mate and in a box bearing no greater proportion to his bulk than a coffin does to a corpse, is inhuman,' wrote one outraged correspondent, deeply offended by the 'cruel spectacle'.

But then, this part of the city was not confined only to animal displays. In his autobiographical *The Prelude*, Wordsworth described the uproarious human zoo of Bartholomew Fair, where, on the land next to the hospital, 'the silver-collared Negro' joined '... Albinos, painted Indians, Dwarfs,/The Horse of knowledge, and the learned

Pig' in a 'Parliament of Monsters'. It was a world out of Blake's progress too, where walls were daubed with apocalyptic graffiti: 'Joanna Southcott', 'Murder Jews' and 'Christ is God'. Set against such scenes, Chunee's death seemed of a piece, conjuring up ancient Rome as much as the imperial city which had replaced it.

In the sedate corner of Lincoln's Inn Fields, the shelved exhibits are stripped bare of flesh, all of a colour, the same pallor you see in your skin when you flex your knuckles. In one jar is the head of a chimpanzee, its eyes half-open like a doll's; in another, a rat drowned in formaldehyde. And in others are human foetuses at every stage of development, floating in tobacco-coloured fluid and pressed-glass wombs.

At the far end of the room is a series of portraits, a decided counterpoint to the worthy surgeons in the hall. They depict piebald children, pathologically obese men, exotic Siamese twins. The saving grace is that they are dignified with names and faces beyond the mere fact of their freakishness, among them Daniel Lambert, whose ballooning body weighed fifty-two stone at his death, and Charles Byrne, whose seven-foot-seven-inch skeleton was acquired by Hunter (at a vast cost of five hundred pounds), despite its owner's desire to be buried at sea. 'The whole tribe of surgeons put in a claim for the poor departed Irish Giant,' one newspaper noted, 'and surrounded his house just as Greenland harpooners would an enormous whale.' Rendered down in Hunter's Earl's Court boiler, Byrne's remains now tower over those of another unfortunate whose crumbling bones desperately tried to regenerate themselves as their owner's contorted spine, ribcage and pelvis grew ever more baroque, sprouting curlicues and coralline osseous splays.

Hunter's three-dimensional index of disease reminds us that pathology is the study of pain itself, from the Greek *pathos*, meaning feeling or suffering. But touring this queasy cabinet of medical curiosities, I'm most taken by the jars which contain the stomach linings of whales, so convoluted that it amazes me my own innards should be so formed, resembling as they do a cavern dense with stalagmites, or the living extrusions of a reef, although later, on another hospital visit, I would watch on a digital screen as a miniature camera snaked up my intestine, revealing its reassuringly pink and worm-like tract, while the consultant and I compared the length of human intestines – as long as a tennis court, he told me – to those of a whale's, which could unroll for a quarter of a mile.

Our bodies are as unknown to us as the ocean, both familiar and strange; the sea inside ourselves. Hunter's assembly of cetacean guts suggests that these creatures and their physiology were equally mysterious to the surgeon – and all the more fascinating for that. 'The animals which inhabit the sea are much less known to us than those found upon the land,' he wrote, 'and the œconomy of those with which we are best acquainted is much less understood; we are therefore too often obliged to reason from analogy where information fails, which must probably *ever continue to be the case*, from our unfitness to pursue our researches in unfathomable waters.' He was the first scientist to describe the cetacea – both inside and out – with any degree of accuracy, and he did so in his paper 'Observations on the structure and œconomy of whales', presented to the Royal Society on 28 June 1787 by Joseph Banks and which, when published in the *Philosophical Transactions of the Royal Society*, ran to more than eighty pages.

Hunter's essay proposed a new equivalence. He sought to show the relationship between form and structure in all living creatures. And what better animal to choose than the whale, one which, as his exhaustive descriptions would show, was like us, and yet far from us too? Hunter would do his best to pursue his researches in unfathomable waters, without ever leaving the land; his cetacean specimens came to him, rather than the other way around.

In 1783 a twenty-one-foot female northern bottlenose whale – *Hyperoodon ampullatus*, a strange, bulbous-headed animal, one of the deep-diving beaked whales – was captured close to London Bridge, the same bridge under which another female of the same species would pass, to certain sensation, more than two hundred years later. This eighteenth-century wanderer, fatally far from home, was acquired by the renowned whale-oil merchant Alderman Pugh, 'who very politely allowed me to examine its structure, and to take away the bones'. Almost as surprising as the sight of Ben Franklin besporting himself in the same waters, the whale's appearance underlined Hunter's frustrations in obtaining specimens. 'Such opportunities too seldom occur, because those animals are only to be found in distant seas, which no one explores in pursuit of natural history; neither can they be brought to us alive from thence, which prevents our receiving their bodies in a state fit for dissection.'

Yet a surprising number of cetaceans ventured into the Thames during Hunter's working life, as though auditioning for his collection. In 1759 a twenty-four-foot grampus (from the contraction of *grande poisson*), otherwise known as the killer whale or orca, was caught at the mouth of the river and brought to Westminster Bridge on a barge. In 1772 another grampus, eighteen feet long, was caught; in 1788 no fewer than seventeen sperm whales stranded on the Thames's lower reaches; and in 1791 a thirty-foot orca was chased up the river as far as Deptford and killed.

Meanwhile, down in Southampton, there were similarly strange forays into the estuary, with similar results. Whales were not unknown here – in the town of Hamwic, bones from Anglo-Saxon whales were used as chopping blocks and worked into combs – and others had passed this way before and since. In May 1770 'a large fish was observed rolling up the river Itchen', and was chased by anglers to Northam, where a nightwatchman saw it strand in the shallows and took the opportunity to attack it with a long knife, stabbing at its head. It proved to be a whale, probably a bottlenose, thirty feet long and six tons in weight. 'There has not been anything of the kind seen in those parts in the memory of man,' claimed the local paper, 'it will therefore be shown at Southam,ton [sic] till the middle of the week.' And in the summer of 1798 the compendious pages of the *Annual Register, or, A View of the History, Politicks, and Literature, of the Year* reported that 'A Fish of enormous size having for several days past been seen swimming in this river, many fruitless attempts were made to take it'; this despite the efforts of Mr Richard Eyamy of the New Forest Rifles Light Dragoons, who managed to lodge a carbine ball in the whale's flanks 'which, it afterwards appeared, went through eighteen inches of solid flesh'.

The animal, also confirmed as a bottlenose, was found the following day languishing on the mud at Marchwood, where three men attacked it, 'forcing an iron crow down its throat, which evidently put it in great torture'. Towed back to the village of Itchen, it was put on public display – 'three-pence each person' – attended by 'an immense concourse' which flocked across the river from the fashionable spa town of Southampton 'to see this uncommon natural curiosity'. These animals, once feared by the Anglo-Saxon author of 'The Whale' – who imagined them swallowing up sailors lured, like fish, into their 'grisly jaws' by their sweet-smelling insides – had become curiosities; but more than that, too. This was the golden age of British whaling, after all, and London was a whaling port, boasting the Greenland Dock, the largest of its kind in the world, surrounded by refineries which supplied oil to light the capital's streets and whalebone to corset the fashionable men and women who strolled them.

In search of fresh and yet more interesting specimens, Hunter commissioned a surgeon – 'at considerable expense', he noted – to sail to the Arctic with a British whaler. Unfortunately, the young man returned with little more than a sample of whale skin covered in parasites. However, Hunter need not have looked so far; the whales came to him, presenting the scientist with a remarkable roster, *viz*:

Of the Delphinis Phocaena, or Porpoise, I have had several, both male and female.

Of the Grampus I have had two; one of them, (Delphinus Orca, Linn. Tab. XLIV.) twenty-four feet long, the belly of a white colour, which terminated at once, the sides and back being black; the other about eighteen feet long, the belly white, but less so than in the former, and shaded off in the dark colour of the black.

Of the Delphinus Delphis, or Bottle-nose Whale (Tab. XLVI.), I had one sent to me by Mr. JENNER, Surgeon, at Berkeley. It was about eleven feet long.

I have also had one twenty-one feet long, resembling this last in the shape of the head, but of a different genus, having only two teeth in the lower jaw (Tab. XLVIL); the belly was white, shaded off into the dark colour of the back. This species is described by DALE in his Antiquities of Harwich. The one which I examined must have been young, for I have a skull of the same kind nearly three times as large, which must have belonged to an animal of thirty or forty feet long.

Of the Balaena rostrata of FABRICIUS, I had one seventeen feet long. (Tab. XLVIII.)

The Balaena Mysticetus, or large Whalebone Whale, the Physter Macrocephalus, or Spermaceti Whale, and the Monodon Monoceros, or Narwhale, have also fallen under my inspection. Some of these I have had opportunities of examining with accuracy, while others I have only examined in part, the animals having been too long kept before I procured them to admit of more than a very superficial inspection.

This fine selection of cetaceans lined up in Hunter's show-rooms, a deceased menagerie to equal the living one at the 'Change. Almost by default Hunter, practised in cutting up human beings, became a whale-dissector extraordinaire. Under the surgeon's knife these whales exposed their inner beauty to the world, the visceral evidence which would tri-umphantly proclaim them *mammals*, not *fish* – a deception in which even Carl Linnaeus had initially been complicit. Pinned to the Swedish professor's door in Uppsala was a cartoonish drawing he had been given of a bottlenose and

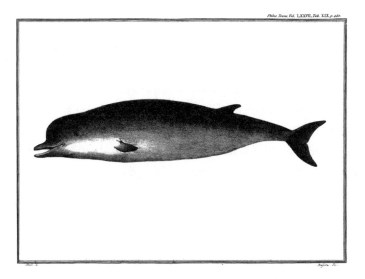

Philos. Trans. Vol. LXXVII, Tab. XIX. p. 460.

calf, a contrast to Hunter's accurate engraving. Its umbilical cord may have convinced Linnaeus otherwise, but the distance between these two depictions spoke of a vast advance which Hunter himself had pioneered.

Hunter dug and delved into the world of the whale. The 'œconomy' of which he wrote was the physical organisation of the animal, one which he was determined to order in an enlightened fashion. Yet the whale withheld as much as it revealed and, almost in spite of himself, Hunter resorted to poetic imagery to encompass its beauty. Faced with a sixty-foot sperm whale, for instance, he noted that the bones of the animal gave little clue to its real shape; that the spaces in between were 'so filled up, as to be altogether concealed, giving the animal externally an uniform and elegant form, resembling an insect enveloped in its chrysalis coat'. Such sleekness merely made these mammals more mysterious, uncluttered as they were by the impedimenta of their land-dwelling cousins with their fingers and toes and feet and hands. The whale was streamlined to its medium, lithe and unencumbered.

But beneath its lustrous layer of blubber, Hunter found that a sperm whale's heart would not fit in a tub; that its tongue was like a feather bed; and that its oil was like butter, 'unctuous to the touch'. The piked whale – which we would call a minke – had five stomachs, as had the porpoise and the grampus, but the bottlenose had seven; Hunter was particularly interested in whale guts. He even observed that, in the interests of science, his sometime pupil Mr Jenner had gone so far as to taste the milk with which the bottlenosed whale suckled its calves, and found it 'rich like Cow's milk to which cream had been added'.

Little escaped Hunter: his report is extraordinarily detailed, a work of beauty in itself, with a sense of deep focus and scientific rigour. He further noted how the nerve endings in a whale's skin indicated a sensitivity which might be even greater in water; and that their eyes appeared adapted to see better in the same medium. Yet his own eye was by no means unerring, as Richard Owen, who would become assistant conservator of the collection, made clear in a later edition of Hunter's works. For instance, when Hunter claimed that the whale's larynx had no function other than in respiration, Owen deferred to his learned friend and co-editor, Thomas Bell, who considered 'the evidence to be strong, if not incontestable, in favour of the existence of a voice in the Cetacea. It is variously described as a bellow, a grunt, or a melancholy cry...'

Those cries remained unheard as Hunter boiled down whales' bodies in his backyard, while copper vats bubbled to prove the purity of spermaceti oil and its point of crystallisation, turning Earl's Court into a whaling station, and its outhouses into ossuaries. Now the remains of these Georgian whales stand arranged in their constituent parts, ready for inspection or even reassembly, should necessity arise; we might genetically re-engineer Eden from these glass shelves and their duly labelled jars. Meanwhile, down in the entrance hall, whale skulls and teeth lie unbiting in dark wooden cabinets, as hippos and rhinos yawn fleshlessly beside them.

John Hunter died in 1793 in mid-argument from a fit of angina, probably induced by his syphilis-weakened heart. Nevertheless his collection, which had been displayed at his house in Leicester Square and which was transferred to the College of Surgeons in 1799, continued to grow. The Hunterian Museum was established in 1811, and expanded in 1855 by Richard Owen. In 1941, three-quarters of the collection was destroyed in a German air raid, smashing into splintered bone and shattered glass the specimens so painstakingly assembled over the centuries, including Chunee's skeleton. Hunter's house in Earl's Court had long since been demolished – having outlived its subsequent use as a private lunatic asylum for ladies, complete with a 'seclusion room' lined with painted canvas. When the site was raked over for redevelopment, bone-filled pits were found, along with evidence of the scientist's experiments into the grafting of trees, their bark excised in the same way he would reduce a limb for amputation; all now covered up by new constructions while lorries thundered along roads which once carried the carcases of whales.

The building is 1930s, brick with white Crittall windows, set back from the street. It might house offices or light industry. Inside, the corridors have the unmistakable smell of an institution. From Rob Deaville's cluttered office – filled with books and equipment and, in one corner, polystyrene containers covered by a sheet of glass under which flies are buzzing in some kind of experiment – I'm led downstairs, past rooms in which young students' faces are lit by screens, into a space that is part changing room, part prison visiting area.

The anteroom is divided into two by a low wooden-slatted bench. One side of the tiled floor, Rob tells me, is 'dirty'; the other is 'clean'. He hands me a well-laundered lab coat, fastened with a row of poppers up one side. As I climb over the bench, I pull a pair of elasticated blue protectors over my boots. My camera, pen and notebook are placed by a sliding glass partition, from where I collect them on the other side.

The laboratory is lined with cabinets containing various instruments. On the far side of the room garage-like doors open out to a view of the trees of Regent's Park. Just over the concrete wall is the zoo and its unseen but vocal inhabitants.

It's a quiet, warm afternoon, the last day in October.

Matt, Rob's colleague, goes through the doors and opens a freezer the size of a small car. From it he pulls a black bin-liner, almost as big as him. It is evidently very heavy. With Rob's help he lifts it to a pair of industrial scales, the thick chains of the pulley system dangling overhead as if in a garage.

I stand, waiting, anxiously. The black plastic bag and its contents are hoisted onto a large stainless-steel dissection table, complete with a sink and drain hole. Rob is talking to me, but I'm not paying attention, because over his shoulder I can see Matt unwrapping the object from the plastic.

From inside the black a slick of brown and red appears. Without ceremony, the subject of this afternoon's study is revealed: a harbour porpoise, *Phocoena phocoena*, a common enough animal around our shores. Until a week ago, this small cetacean was swimming around Cardigan Bay, feeding on small fish. Now it lies in a basement of the Zoological Society of London, quite calmly, almost as if asleep.

The fluorescent light is reflected onto its flanks by the dull sheen of the stainless steel. It is subtly marked, brown with greyish striations. Its tail is elegant, on a small scale; its dorsal stubby. The animal is plump. In fact, it looks remarkably healthy, and one might almost imagine it could swim away – if not for the fact that its eyes have been eaten out of its sockets, and its side has been pecked by gulls, leaving dents like those made by a pencil point in an eraser. And when I walk around to the other side, I discover that one half of its face has been entirely eaten away. Yet the carcase – probably about four and a half feet long from its curved flukes to its snub head – has the dignity which all dead things have, from birds to humans. Death wipes away fear, leaving beauty behind. I feel guilty as I photograph it, invading its privacy; an animal out of its element and laid on cold metal, instead of being suspended in water. The images I take are forensic; I suppose that's what my body looked like when I was photographed for medical science. I take one last look at the animal, whose wholeness is about to be destroyed.

Rob talks me through the cetacean's dimensions, and what he is about to do. Picking up the scalpel onto whose handle he has just snapped a new blade, he cuts through the flank with deft and unerring swiftness. The dark skin opens to show the glaring white beneath – the colour of coconut. With an adeptness a sushi chef would envy, Rob slices out a long sliver of the blubber. He lays out the section like a bacon rasher, and carefully measures it. 'It's a healthy layer,' he says. 'It gets thicker in the winter, to keep the animal warm.' Already, he has spotted something anomalous: a cavity in the blubber, filled with blood, big enough for Rob to poke his finger in. It is the first internal sign that this animal did not die of natural causes.

'See – the ribs are snapped, here, and here,' says Rob, whose own movements are, ironically, compromised by over-adventurous paragliding at the weekend which has left him with cracked ribs. Matt is summoned to cut out the ribcage with a long-handled pair of cutters. The bones are loosened and laid out, like spare ribs, sticky with sweet and sour sauce. 'Porpoise meat is good eating, you know,' as Ishmael says.

Turning his attention to the animal's underside, Rob removes two identical organs from its interior. They resemble a pair of elongated plums: the porpoise's testes. Given the size of the animal, they're remarkably large, I say. 'It's because they're so sexually active – like dolphins,' Rob says. 'The testes of a dolphin can weigh two kilograms.' There they lie, by the side of the porpoise. No use to him now.

Slowly Rob works his way through the rest of the animal, carefully excising each organ. It's like taking apart a three-dimensional, bloody jigsaw. The rest of the blubber is peeled off in large white slabs and tossed into a large yellow plastic sack. CITES, the Convention on International Trade in Endangered Species, demands that every part of a protected animal such as this one – even though there may be half a million porpoises in British waters – must be bagged up and taken away to be incinerated. Matt stands by with small plastic containers for the tissue specimens Rob saves for later analysis.

Soon the steel is awash with blood, its surface strewn with offal like a miniature whaling station. Even Rob and Matt admit to each other, after a beer or two, an occasional sense of revulsion at their work, although it reveals something miraculous: the essential secrets of the cetacean, so like us that it might be a child lying there on the slab, deconstructed and, bit by bit, dumped in the bin. I realise that the animal is as beautiful inside as it is outside; I ought to put down my camera and paint it instead. Digital images only heighten the lurid quality, and lose the subtle colouring of what lies inside; nor can they convey the faint but evocative smell of the sea that still hangs about the carcase.

The largest organ of all, lying across the others, is the liver. As Rob cuts into it – its interior still crackling with ice like a partly-defrosted steak – he discovers a huge tear through its centre. One almighty blow caused this rupture, and the congealed blood that lies around it. There are more contusions on the other side of the animal; clearly it was the victim of a sustained attack.

Out of the chambers of its lungs, Rob's fingers tease slender, long worms that infest the organs. It's hard to resist a sense of disgust at this discovery, even though this is a minor infestation compared to the thousands of nematode worms found in large whales. Rob tells me about a particular parasite that gravitates to the genital regions of small cetaceans, from where they can complete the final cycle in their parasitic lives. There a shark will nip and gnaw at its victim, and in the act of ingestion provide the parasite with its end-user, its ultimate destination.

As I watch Rob pull a two-inch worm out of the porpoise's lungs, I ask if they harm the animal. Rob says they don't appear to; but there are areas of calcification in the lungs, too, hard and brittle within the surrounding tissue, a result of former infections. It may have looked healthy, but there were plenty of assaults on this porpoise's well-being; the presence of anthropogenic contaminants may reduce cetaceans' resistance to such infections.

Taking hold of the tongue, Rob removes the 'pluck' of the animal – the mouth, oesophagus and heart. They come away in one long lump. The heart is healthy, and looks as though it has only just stopped pumping blood around the animal's body. Rob pulls out the valves that did this sterling work, then it too is consigned to the bin, just another piece of offal, having served its useful purpose.

Now all that is left is the head, lying forlornly in a corner of the shallow sink. The mouth is agape. Its two rows of tiny teeth are rounded, rather than pointed like those of a dolphin, designed to grabble about in the sea bed, searching for the small prey that constitute most of the porpoise's diet. Earlier, in its stomach folds, we found tiny squid beaks and fish otoliths, minute ear bones.

Rob turns the head on its side and locates a tiny dent in the skin behind the eye. You need to know a cetacean inside-out to be able to locate the external aperture of this organ. The hole is barely more than a pinprick. Rob's scalpel swiftly slices it open to show the thread-like auditory canal, a dark root running down to the tympanum. It is a bare, minimal form, since most of what the animal hears is conveyed by the spongy, oily tissue that lines its jaw.

This strange cetaceous sound equipment – a scaled-down version of what one would find in larger-toothed whales – is thus exposed. We follow it, with the aid of Rob's knife, from the 'melon' or forehead, a thick white wall of fat with bio-acoustical properties, to the 'monkey's lips' buried deep below, a cartilaginous valve that snaps together to create the animal's clicks.

Finally, Rob digs out the ear bones or ossicles, so dense that they're the last part of any cetacean to survive. They're exquisitely shaped, like shells. I have three in my own collection: a time-darkened, fifteen-million-year-old fossilised specimen as heavy as a stone; a yellowing otolith from a pilot whale which had been butchered sometime in the early twentieth century; and a fresh sculptural shape I found when walking on a remote beach, bone-pale and light and multi-chambered, so intricate as to be a miracle in itself. These convoluted objects, more like musical instruments, are all that remains of their owners, their existence reduced to the echo chambers of their most intense sense. They hold the last sound they heard, just as our hearing is the last we will know of the world.

Matt unplugs a battery-powered circular saw from its charger on the wall and hands it to his colleague; the noise revs through the room. Rob tells me, 'You may want to step outside while I do this – there's a possibility of aerosol.' As he pulls down the heavy Perspex visor over his head, I move into the welcome fresh air. I have already been instructed not to put my hands in my pockets or touch my lab coat with my face, for fear of zoonotic disease; with an animal so genetically close to my own species, infection is a real risk.

Outside the room – filled as it is now with the mingled smell of blood and the sea – the zoo's inhabitants are going about their business, unaware of what is happening to one of their fellow creatures. I imagine other subjects that have passed through this room – from a giant leatherback turtle splayed over the tiles, to other exotic animals from the modern menagerie next door: elephants, antelopes, apes.

The high-pitched squeal of machinery stops, and Rob calls me back in. He has cut off the back of the creature's skull. There's a burning smell in the air, familiar from the dentist's chair. White flecks of sawn bone are strewn over the red tissue. Rob eases away the section of cranium and digs around inside with his scalpel, as though working on a giant oyster. With a plop, the loosened brain falls out into his hands. It fills his cupped palms like a wobbly, incarnadined blancmange. Jiggling it in his fingers, Rob points out the cranial lobes and the cerebellum. What information did this organ process in the final few minutes of its owner's existence? Earlier in the dissection, we'd seen the porpoise's overdeveloped adrenal glands, bigger in proportion because of all the stress the animal had suffered in its life.

It is now that Rob reveals the reason for this animal's death; a conclusion he suspected all along, and which its blood-flooded brain confirms. The porpoise was murdered by its own cousins, the bottlenose dolphins of Cardigan Bay, the latest in a long line of such fatalities, no fewer than three hundred known incidents in the past twenty years, perhaps indicating many more.

I imagine that attack. Some testosterone-fuelled young dolphin decided to play with its fellow cetacean's life, treating its body like a rugby ball, tossing it high in the air with a flick of its powerful hard beak, then repeatedly butting it till the porpoise's ribs snapped and its liver split. I hope that it lost consciousness before the gulls descended to peck out its eyes. Later, I speak to a researcher from the scene of the crime. She tells me she has often found dead porpoises on the beach that look similarly unscathed, until, with a depression of the foot, you realise every rib has been broken. The phenomenon may be a recent one, or it may not. Perhaps the dolphins' ever-present smiles have kept us from this shocking discovery: that they could be such cold-hearted killers. Mostly young adult males are to blame, and attacks happen around periods of mating – indicating that the porpoises may be the victims of male aggression over access rights to fertile females.

Adult males may mistake the porpoises for infant dolphins whose existence challenges their genetic legacy. Whatever the reason, the evidence is clear: dolphins are not the benevolent mammals we'd like them to be; those beaming faces hide the minds of assassins. A few months later, I'd visit Spey Bay in the Highlands of Scotland, home of another resident pod of bottlenose dolphins; they too have been implicated in offences against their cousins. As they swim under the boat, out of the misty North Sea, I realise that they are huge creatures, up to twelve feet long, almost black, with white bellies that flash as they leap. It's easy to imagine their power; or why one local man tells me that when he goes fishing up to his thighs in the bay and sees those dark fins coming nearer, he quickly orders his dog out of the water, for fear – rational or not – that they might turn and go for it.

As I stand in the dissection room, looking at this carcase laid out for my edification, I feel a sense of privilege to have been granted such an access to its inner secrets. Its anatomy is so close to ours; every excised organ is a reminder that what is contained within the mammal's blubbery coat is also held together by my bag of skin and frame of bones. As it is taken apart, so am I, all my bones and organs and skin and guts. I'm looking at myself.

Outside, the macaws are calling loudly from their cages. I cycle off into the dusk, the smell of porpoise in my hair.

4
The azure sea

These whales are reported to have their lairs...their own territories and apportioned dwelling-places, and remain there without trespassing on their neighbours' preserves. They do not rove aimlessly to and fro seeking changes of abode, but love their own home as if it were their native country, and find it gratifying to linger there.

St Ambrose, *Hexameron*

From my plane window, the islands seem to rise up as if newly erupted from the sea. It's four years to the day since I last came here. I know that because the same festival is in progress. Perhaps it never ended.

Every so often rockets fizz into the air. Troupes of children and adults are dancing and singing, each followed by their own brass band. Teenage boys who elsewhere would be embarrassed to take part in such a procession are dressed in satin bows. They and their partners pirouette along the route, dancing down to the square, where, the following day, in the shadow of the basalt-outlined church, tables will be loaded with round loaves of sweet bread, stuck over with flowers and offered up to São Pedro, patron saint of fishermen. For all these joyous celebrations, however, a current of reservation runs through these streets, defined as they are by the sea.

A few months before that last visit my mother had died. My grief seemed implicit in this place. I felt open to its remoteness and obscurity, its clinker-dark shores which looked as though the whole place had burnt to the ground.

The mid-Atlantic night is pitch-black and impenetrable. The moonless sky sinks into the sea, allowing the Cory's shearwaters to sail inland to their nests, feathery ghosts in the darkness. All day they've worked the waters, seizing fish and squid from below the waves. But as evening falls, they burst into eerie squeaks – *qwwwaaark—qwak-qwak—qwwwaaark* – strangulated, semi-human, half-comical cries constantly reiterated as they soar inshore, each sounding more demonic than the last; little wonder that their calls were once thought to be those of the devil.

They owe their common name to Charles B. Cory, the nineteenth-century ornithologist and golfer who first described them in 1881, and whose house in Boston was stuffed with nineteen thousand avian specimens; their Latin title is rather more prolix, *Calonectris diomedea borealis*. Shearwaters – which do precisely that – hail from the family of procellarids, after *procella* or storm, the kind of weather they are supposed to favour. It was one reason why sailors were wary of their appearance; another being that they were believed to bear the souls of their drowned comrades.

Their order includes some of the most romantic birds of the sea: the albatrosses, fulmars, puffins and storm petrels. Like whales, they have evolved mechanisms to cope with an oceanic life: olfactory bulbs in their brains that allow them to smell their food from far away; tubes in their beaks that discharge salt water; and a diet of fish, squid and krill which produces a noisome stomach oil potent enough to dissolve the plumage of any interloper who is on the receiving end of a projectile vomit, although its greatest benefit is as a highly calorific fuel. Procellarids require such energy, since they are the greatest riders of oceanic winds, which echo the currents below. From wandering albatrosses with twelve-foot wingspans to tiny storm petrels barely bigger than a sparrow, they undertake journeys so ambitious that they occasionally get lost and end up in strange locations, such as the back of a whale. Or perhaps not so strange, since their kind have long been called 'whale birds' after the way they appear to herald leviathans. Storm petrels will fly downwind of hunting orcas, savouring the fishy grease in their blows. Off Cape Cod, I've watched clouds of greater shearwaters over feeding humpbacks, even dipping into the whales' mouths to pluck out sand eels.

In the waters off the Azores, whales and dolphins drive to the surface the deep-water prey that the birds cannot ordinarily reach; as the cetaceans round up their bait, the shearwaters dive down to feed on the same source. Sadly, these seabirds suffer from our own hunting habits. Tens of thousands of procellarids die each year snared on longlines, strung out along them like a gamekeeper's line; some escape only to return to their nests, beaks loaded and pierced with hooks.

Cory's shearwaters mate for life, returning to the same site each year to lay a single egg. They regurgitate their day's catch, supplemented with rich stomach oil, into the mouths of their begging chicks, all under the cover of a dark moon to avoid predatory gulls. With their legs at the back of their bodies, setting their centre of gravity awry

167

like wonky clockwork toys, they are at their most vulnerable on the ground. Yet these islands must suit such precarious lives, since they are home to seventy thousand breeding pairs of shearwaters that gather here from all over the North Atlantic, very noisily. They echo in my ears as I lose consciousness, and rouse me a few hours later when the birds start to leave before dawn. Too late to sleep, too soon to rise, I give in, lie awake, and listen.

I realise that each cry is individual (odd how we assume all animals of a species sound the same, as if everyone we knew spoke in an identical intonation). I start to hear how the intervals between each screech change; how they elide the two central notes of their four-note phrase, and how the final squawk goes up in tone. I practise Cory calls in my head, wondering if they're different from the sounds they make as they come in – 'Don't worry, I'm here, I'm home,' as opposed to 'See you later, don't worry, I'll be back soon.' Like other nesting seabirds such as the gannet, their mates can recognise each other's cries over the hubbub, a phenomenon known as 'the cocktail-party effect'. There is also sexual dimorphism at work, a behavioural difference between the genders: the male's call is distinctly more ringing in tone, presumably to attract its mate.

Mulling over these sounds in my head and what they may or may not mean, I drift back to sleep till the sun rises over the volcano and it's time to go back to sea myself.

I'd forgotten how uncomfortable it was – the back-breaking, ball-numbing ride as the hull hits the waves with a mighty thwack. I wonder if our skipper enjoys inflicting such pain, but João, who is also a footballer, complains of the effect of his day job on his spine. Every rucking wave means another bone-jarring bump – *pow-pow-pow* – every fall harder for each rise. It's the price we must pay for our presumption, for daring to enter their domain. I decide that it's better to stand up, legs astride the padded seat like a cowboy at a rodeo. It works, after a fashion, but I keep checking my teeth to see that they're still intact.

As suddenly as it kicked into life, the boat judders to a halt as João cuts off the engines. Over the Tannoy, we hear a quiet voice speaking in Azorean Portuguese. Up on the cliff top, perched on his wooden stool in the Vigia da Queimada, Marcelo André da Silva Soares is calmly directing our movements.

Marcelo is twenty-four years old, and has lived with whales all his life; his father watched them from the same concrete tower on the headland, directing his fellow hunters. In the transition from hunting to watching, which happened barely a generation ago, Marcelo's father passed on his knowledge to his son. Now, for ten or twelve hours a day, Marcelo sits quietly in the vigia, peering through powerful binoculars. Behind him is a chart of different species and shapes of cetaceans, resembling the outlines of aeroplanes provided to wartime spotters. From this elevated position, Marcelo's field of vision can range over two dozen miles across an ocean tilted up to the land like a board on which every movement may be plotted. He is a grandmaster, strategically moving his pieces – the boats – to their targets – the whales.

Marcelo's father still works a neighbouring vigia, too. Where his voice is urgent, with the excited stridency of a market trader, his son's words are whispered, in the hushed tones of a late-night radio DJ. He may be half my age, but Marcelo has already lived another life as a soldier in Kosovo, where other hunters – snipers with deadly rifles – also took their time, and their aim.

Marcelo directs João towards a whale which has just surfaced. Both men know their part. They know this is a narrow window of opportunity. The whale will stay up for less than ten minutes, as long as it needs to charge its body with oxygen before diving again. And in the time it takes us to get there, the animal promptly does so, leaving only the distant glimpse of its descending flukes and misty blow.

From behind his plexiglass screen, João unwinds a wire and tosses a hydrophone into the sea. Immediately, we hear clicks over the loudspeaker: the steady, insistent sound of whales in search mode, scanning the water column for squid: a sharp stutter, a pure, clean sound compared to the messy squawk of the shearwaters. It reminds me of the hundreds of bats that skittered over our heads last night, roused from their roosts and employing the same technology.

A sperm whale's clicks are the product of a process even more complex than Southampton's double tides. They're created by the opening and shutting of a valve set at the front of the animal's head, the same device found on porpoises and dolphins. Known as the *museau de singe* or phonic lips, they operate rather like a human voice-box, sending the rapid sounds back through the whale's internal, right nasal passage and through its 'upper case', a reservoir of oil with bio-acoustical properties.

Having travelled all the way to the back of the whale's concave cranium, the sound then bounces from an air sac, then returns through the 'lower case', filled with convex compartments of spermaceti oil which act as acoustic lenses. Only then, having passed from the front to the back of the whale's head and back to the front again, is this powerful, refined sound emitted from its nose and out into the ocean. Then the whole procedure begins again (and all this at rates of a hundred times a minute). The highly directional sonar pulse bounces back from its target to be received, not through the whale's outer ear, which would barely admit my little finger, but through its huge tuning-fork jaw. From there it is conducted to its inner ear. It has even been suggested that the rows of teeth along a sperm whale's jaw act as regulators of this industrial-scale sound system.

Reduced rather than amplified through João's puny loudspeaker, the whales' clicking scans the ocean a thousand metres below. It is the sound of control, at its peak, the loudest noise made by any animal, louder than a jet engine; but it is also entirely subtle, obeying its own laws of sequence, timbre and rhythm.

First come the chattering 'codas' as the whales appear to discuss what they are about to do, or tell a waiting calf, 'I'll be right back' – not unlike the shearwaters. Then follows a regular pulse, scoping out the ocean bed, sizing up the field, as it were. Next come the regular clicks that indicate a diving whale is searching for prey. If it is successful, it will emit a series of accelerated buzzes that mean it has locked onto something to eat, and on which it may still be feeding as it returns to the surface to resume its social codas. These may mean nothing more than 'I'm back,' but they could have more complex, conversational associations. These are, after all, animals with the biggest brains in nature, with a matrilineal culture reaching back for millennia. They are as different from other whales as ravens are from other birds; *über*-whales, a species set apart.

If we pay attention, says João, we can hear the rhythms of individual whales, as characteristic as a Cory's cry. 'Some are more metallic than others – it's their different voices,' he says, in his own clipped fashion. This is the cocktail-party effect on a gigantic scale, but it doesn't always do to shout. 'At the surface, they're silent. They don't want to be heard by predators. And since they can see here, they don't need their sonar.'

Together with Yuri, João's first mate – a tall French boy with cropped dark hair – we discuss the naming of the whale. In French, as in other Latin languages, the sperm whale is known as the *cachalot*; Yuri points out that this has a double meaning: it sounds like *cacher à l'eau* – to hide in the water. To speak is to draw attention to yourself, and beyond the bounds of your tribe attention is seldom welcome. That's why the whales are silent at the surface, where they come closest to us: in silence they might also be invisible, these hiders in the sea.

Yet shouting is what whales do best. They live in an element in which noise travels five times faster than in air. Their brains are wired for sound; their auditory cortex is larger than our visual cortex. Such a capacity is essential for animals that hunt in the lightless depths. Theirs is a very different experience of the world from ours, because their world is so different. For toothed whales blessed with pin-sharp sonar accuracy, everything is transparent; nothing is concealed. They live in another dimension, able to see into and through the solid, to discern structures inside. A whale or a dolphin can see the interior of my body as accurately as I can see the exterior of hers; I must resemble one of the educational models we had at school, clear plastic figurines of a man and a woman with their organs indecently displayed. The world is naked to a cetacean.

In his book *In Defense of Dolphins*, Thomas I. White, a professor of ethics, notes that his subjects are able to use their sonar to detect one another's emotional states by the way their temperature falls or rises, like a human lie-detector test. As a result they cannot dissemble about the

173

way they feel, as we do. They know if another dolphin is angry or excited. Citing the dolphin's enlarged amygdala, the primary processing centre of the brain that deals with emotional and social connections and which is proportionally much larger in dolphins than in humans, White suggests that they may actually be more emotionally developed than us, partly as a result of their high level of social activity: they need to get along with one another since they travel in such close proximity and great numbers. This may be even more important among sperm whales, as Hal Whitehead, the eminent cetologist, says, since at any time they could turn their powerful sonar on their neighbour and cause serious damage. Whales must have codes of etiquette, perhaps even morality. Good manners may be as desirable in cetacean gatherings as they are in ours.

For humans, emotion may be merely a product of our evolved brains, a function of the spindle cells that distinguish us from other mammals. But it has recently been discovered that some cetaceans – including sperm whales – also possess these bundles of nerves, up to three times as many as humans and primates, and evolved them thirty million years ago, twice as long ago as us. Such cells process social organisation, speech, and intuition about the feelings of others; as to what use whales put these emotions, or even if they are anything like our own, we simply do not know, and can only imagine. If it is empathy that marks us out from other animals, what if that same sense of fellow feeling exists among cetaceans? Could that be a reason for their predilection to strand en masse? Hal Whitehead relates an incident in which a single male

174

sperm whale beached itself on a remote shore, while two of its fellow whales swam up and down the bay, becoming ever more distressed. Eventually they too stranded themselves, to die alongside their comrade.

It is another refinement of our humanity that, unlike other animals, we know we are going to die. Research has shown that cetaceans have a sense of individual self, and are aware of themselves as sentient creatures. What if they shared our existential angst, too? We are unable to interview animals; they leave no autobiographies. Despite our science, their interior lives remain a mystery.

Below us, the steady searching clicks continue. João listens some more. Then, satisfied with what he's heard, he predicts the reappearance of 'our' whale as his speaker falls silent. Has the hunter found its lunch? Is it even now sucking in squid through its narrow, tubular mouth, swinging open its slender jaw, glistening white, as if to lure them in? Whether it got lucky or not, the whale is ready to return to our world. And we will be ready for it.

Up on the rocky headland that overlooks this spectacle, I sit alongside Marcelo on a wooden stool in his concrete cell. Silently, we watch the waves unroll. A few moments before, from the top of the field behind the vigia, up the grassy path where lizards scatter with every step, I had spotted dark shapes in the sea, moving swiftly. Five or six of them, long and sleek, with dorsal fins set far back. Dashing back to the vigia, I consulted with Marcelo and his charts. I realised there's only one type of whale that matches this fast-moving group – which has now vanished into the silvery expanse and is nowhere to be seen.

Beaked whales are bizarre animals. Some have heads that resemble a bird's, while their bodies are spindle-shaped, more like archeocetes, the ancient whales; their sharp beaks evoke even older ichthyosaurs. In 1823, when the French naturalist Georges Cuvier discovered the skull of the beaked whale that would bear his name, he assumed the animal to which it belonged was extinct. It took fifty years for scientists to establish that the species was still alive.

More than any other whale, the family of Ziphiidae elude our scrutiny, swimming in the deep ocean, far from land. Later, in New Zealand, Anton van Helden, a world expert on beaked whales, would show me the skull of a spade-toothed whale, *Mesoplodon traversii*. The cranium, a scooped-out ski-slope of calciated matter, more abstract sculpture than bone, its lower jaw studded with what look like stumpy tusks, stands on a shelf in a storeroom of the Te Papa museum in Wellington, until now one of only three known specimens; it is the world's rarest whale, and has never been seen alive. Only recently a cow/calf pair was washed ashore on the Bay of Plenty, giving the first true idea of what these five-metre-long animals look like. In the past two decades alone, three new species of beaked whales have been identified, raising the total to twenty-one, although they're constantly being revised – six exotic genera in search of themselves, with names as strange as the animals they evoke: *Ziphius*, *Tasmacetus*, *Berardius*, *Indopacetus*, *Hyperoodon*, *Mesoplodon*.

One reason for such obscurity is that beaked whales spend so much of their time below the surface, foraging in the depths where they suck squid through their mouths. So seldom seen, they exist in a category of their own. They are (mostly) defined by their prominent beaks, often with a pair of teeth that jut out even when their mouths are shut. Their slender bodies are perfectly suited to diving; their pectoral fins fit into 'pockets' to reduce drag. They've found their evolutionary niche, these middling-sized whales; one lying on my garage roof would just about overhang it. As varied as any family of songbirds, they suggest some avian-cetacean hybrid, especially the Cuvier's beaked whale, *Ziphius cavirostris,* which I once saw from the unlikely vantage point of a cross-Channel ferry.

As we approached northern Spain over the deep underwater shelf of the Trevelyan escarpment, which plunges to four thousand feet and thence towards the Azorean plateau, three brownish shapes appeared. Through my binoculars I could see the scratches on their backs, the results of battles with one another and with squid; one even breached in front of the prow. But more striking than anything – and more tantalising – were their heads, oddly pinkish as if veiled with an amniotic caul, and distinctly bird-like in shape. It was easy to see why its common name is the goose-beaked whale, a chimerical confusion that invokes tales of barnacle geese born of molluscs.

Like the Cuvier's, other common names commemorate a list of long-dead scientists: Arnoux's beaked whale, Andrew's beaked whale, Longman's beaked whale, Hector's beaked whale, Gray's beaked whale, Baird's beaked whale, Blainville's beaked whale, True's beaked whale. Other, equally cumbersome denominations amplify their owners' dentition: the strap-toothed whale, the spade-toothed whale, the gingko-toothed whale. A cetacean orthodontist would surely shake his head and say, 'What can I do with teeth like that?'

Yet their owners surely rejoice in such splendid canines. On Blainville's beaked whale, as on others, the teeth are resplendent with purple-stalked barnacles as exuberant, weedy adornments dangling from their mouths. Like a deer's antlers, such tusk-like teeth may be used as a secondary sexual signal, enabling females to sort out males of their own species, since even to beaked whales, other beaked whales look the same. But the animals themselves are, on closer inspection, subtly coloured, a muddy range of greys and browns and blacks, striated and crisscrossed by innumerable scratches, as if subjected to cosmic strikes. Each might be a heavenly body, like Saturn's icy moon Europa with its 'chaotic' terrain cracked and riven by an unknown ocean. (Indeed, astro-biologists speculate that the exosolar planet Kepler-22b could be a watery world where whale-like animals swim through atmospheres of sea; the word planet, after all, means 'wandering star'.)

Like those remote astral bodies, which may be detected only when they dim the light of their parent suns almost imperceptibly as they pass between them and us, it is extraordinarily difficult to discern these strange species at sea, even for the most experienced cetologist. They remain a remarkable absence. Melville omits them from his Cetology, but given his love of the eccentric and the paradoxical, I dare say he would have had words for them. To me, they appear to be engaged in a cryptic choreography of their own, out there in the oceanic universe, a masked ball of beaked whales, elegantly pirouetting beyond the human gaze. But then, I probably think too much about whales, generally.

The next day, after my visit to Marcelo in the vigia, reports come in confirming a group of Sowerby's beaked whales, *Mesoplon bidens*, first identified in 1804 by the English naturalist James Sowerby. As social, toothed whales, beaked whales are particularly prone to stranding, and increasingly, it seems, susceptible to human-generated sonar because of their highly attuned reliance on sound. In one well-documented case in 2002, a mass stranding of Cuvier's beaked whales occurred on the Canary Islands, four hours after the onset of military exercises in the area. Necropsies of the animals revealed the kind of haemorrhages and gas-bubble lesions associated with a build-up of nitrogen in the blood, known as necrosis – what human divers call 'the bends'. The whales seemed to have been panicked by the noise into surfacing too early; other reports indicate that similarly frightened whales adopt an unnatural up-and-down 'flight mode'.

Such strandings address us directly, as emblems of our careless actions. In *The Sea, The Sea*, Charles Arrowby's cousin James informs him, 'The sea is not all that clean... Did you know that dolphins sometimes commit suicide by leaping onto the land because they're so tormented by parasites?' To which Charles replies, 'I wish you hadn't told me that. Dolphins are such good beasts. So even they have their attendant demons.' In recent mass strandings of common dolphin on Cape Cod, rescuers discovered that the animals became less stressed if they were placed side by side, even as they lay beached and gasping on the sand. In their distress, their only consolation was each other.

On 11 December 2009 there was a remarkable stranding on a southern Italian beach in the Mediterranean, where a 'lost tribe' of sperm whales live, distinct from their oceanic cousins, isolated in an inland sea. Their fate had all the elements of a classical tragedy. The seven whales, all males between fifteen and twenty-five years in age and measuring ten to thirteen metres long, had been driven into shallow waters, possibly by military sonar. Unable to forage on deep-sea squid, they first began to dehydrate, then to starve.

Now a third threat came into play. In their state of hunger, their bodies began to break down their adipose fat, releasing the heavy metals and organochlorines they had inadvertently ingested from the polluted seas, and which had been absorbed into their bloodstream. In effect, the whales were poisoning themselves. As Dr John Wise showed me, in his laboratory at the University of Southern Maine, Portland, sperm whales are particularly susceptible to pollutants, not only because of their position at the top of the food chain, but due to the way they breathe so deeply. While I watched sperm-whale cells multiplying in a Petri dish under one of his electron microscopes, Dr Wise told me how sperm whales, which range widely in their oceanic travels, inhale chromium emitted from coastal chemical plants which, along with other contaminants, may be causing cancer and birth defects analogous to Down's syndrome in human beings. We share the same air as mammals, yet we contrive to poison even that, as if not content with bespoiling the sea. The same chemicals that created Rachel Carson's silent spring might yet silence the world of the whales.

For the ill-fated spermaceti septet in the Mediterranean, that toxic cocktail weakened their bodies yet further, altering their sense of orientation and perception. The fishing gear, hooks, rope and plastic objects subsequently found in the whales' stomachs – a result of their choosing to live in an inland sea subject to so much human detritus – hardly helped. (Recently, a juvenile sperm whale was found floating dead off Mykonos. The animal was emaciated, yet its stomach was distended; when it was cut open, nearly one hundred supermarket bags and other bits of plastic debris spilled out.)

The last unlucky component in these, the last few hours of their unlucky lives, was the stormy weather that conspired to drive the seven whales inshore – possibly following the first of their number to give up and thereby demonstrating the loyalty for which their species is renowned. Four of the whales had already expired by the time they were discovered on the beach. The other three took days to die, ultimately suffocating under their own enormous weight; that which had sustained them at sea, a marker of their success as animals, doomed them on land. It must have been a painful death, reflected even in the usually measured words of the scientific paper, which described the whales 'found agonizing on the shore'. In such circumstances, we humans look on helplessly. Smaller cetaceans can be given an overdose of horse tranquilliser which swiftly inhibits respiration and causes death. But for these leviathans, there is no such mercy. 'It would be impossible to get enough quantities of the drug to euthanise a sperm whale,' says Rob Deaville, 'and even if we could, I doubt we could inject it.'

The theories surrounding strandings are dizzying in their claims and counterclaims. Some focus on the whales' ability to follow geomagnetic lines laid down in the earth's crust. Such sensitivity – via minute magnetic cells that spin like internal compasses – has been detected in organisms from bacteria to birds. Birds in particular are thought to possess photopigments in their eyes known as cryptochromes that detect the magnetic field chemically, seeing it as a pattern of colours or lights which enables them to navigate. Could whales 'see' these same patterns? Some studies of the British coast, where the geomagnetic contour lines run parallel to the land, have suggested that whales move along 'geomagnetic valleys', and that where such valleys lead inland, strandings may occur.

Others speculate that cetaceans set their 'travel clocks' by detecting these minute changes in the geomagnetic field; or that the circumstances for strandings may be created by sunspots known to affect the earth's magnetic field, most visibly in the aurorae borealis and australis. Even more radical hypotheses suggest that cetaceans display a foreknowledge of seismic shifts in the earth's surface, as if they were canaries warning of disasters as yet undetected by humans. The fact that recent major earthquakes in Japan and New Zealand were preceded by mass strandings of pilot and melon-headed whales seems to support this notion. But if we were ever able to lock into the magnetic fields that surround us, and that guide storm petrels and sperm whales alike, we lost the ability long ago. Our senses are sadly lacking, even in the three dimensions we purport to perceive.

Out in the mid-Atlantic, we spend hours searching for sperm whales. 'They're acting weird today,' says João, 'playing games with us.' We're about to turn back when a pod of Risso's dolphins appears out of nowhere.

I've only ever seen these animals from afar. Now they're just off the bow. Scarred and scratched, they resemble damaged ghosts, caught out of the corner of the eye. They even behave differently from other dolphins, staying shyly below the surface. Only as they come closer can I see their blunt snouts and high dorsals, their flanks graphically black and white, like psychedelic zebras. Antoine Risso, a French contemporary of John Hunter's with a particular interest in crustaceans and copepods, lent his name to these animals. In the Azores they are known as *moleiro*, for their whiteness.

Sliding over the side of the boat, I see their shapes moving below me in the gloom. Through the water I hear them singing – a sweet, high-pitched song rising up from these cetacean choristers in the sea's cathedral. Melville does include them in his Cetology, although whether he ever saw one is uncertain: 'Though this fish, whose loud sonorous breathing, or rather blowing, has furnished a proverb to landsmen, is so well known a denizen of the deep, yet is he not popularly classed among whales. But possessing all the grand distinctive features of the leviathan, most naturalists have recognised him for one... By some fishermen his approach is regarded as premonitory of the advance of the great sperm whale.'

As they rise to the surface, their enormous dorsals appear, enough to identify them even without a glimpse of their battle-scarred backs, which resemble those of beaked whales. I peer at them through my mask; to see these cetaceans so close and yet so elusive only makes them seem more particular.

Back on shore, Karin Hartman, a Dutch scientist studying Risso's dolphins, tells us a little more. She says the males are whiter than the females, partly because they fight more and find more squid, leaving their scarred skin unpigmented. The paler they are, the more attractive they are to the opposite sex, as good foragers and representatives of a tough and frisky breed. 'It's sexy to be white,' says Karin. The whitest are also the oldest, since their skin becomes thinner as they age.

As with other whales, this ratio of dark and light is an advertisement of their individuality. Yet like sperm whales and beaked whales, their teeth are not used for feeding; they employ suction to feed on the squid of which they are so inordinately fond. Can there be enough calamari in the sea for these creatures? Hal Whitehead tells me sperm whales eat one hundred million tons of fish and squid each year, more than we humans take out of the oceans.*

I fall back in the water, into a flurry of fins and limbs. We're caught up in a trio of sperm whales, almost squashed in between them. Their big square heads float past mine, eyes and flanks, a confusion of us and them. It takes moments to sort us out – cetacean from human – before the whales dive, leaving my diving partner Drew and me to haul ourselves back on the boat.

We sit squeakily on the rubbery side of the rib, awaiting our orders. Another blow appears close to our bow. João manoeuvres alongside the animal, slowly closing the distance between us. He tells us to be as quiet as we can. Not for the first time, it occurs to me how odd it is that an animal ten times the size of the craft on which we sit should be so timid of our proximity; as though, like the wheatear, it might be frightened by clouds, which is what we represent – a black rubber cloud over the whale's head.

My natural reaction in the water, to reach out and pull myself through it, is inappropriate with a wild animal; I might as well be waving my arms in front of a hippopotamus. Drew shows me how to drift back with the boat as we drop in, minimising the disturbance; how to keep my fins below the surface, so as not to create a stream of bubbles – a sign of aggression to a whale.

Everything is about making our bodies as unintimidating as possible. At barely five foot eight and eight and

* *Note to editor.* Maybe we shouldn't publicise this? Imagine the protests fishermen might lodge about these greedy leviathans. I doubt that they'll be reassured by the argument that the whales were here first.

a half stone, how could I present any peril to an animal with a body mass so many times my own? Yet even before I get in the water, I'm inflicting stress on a creature whose well-being I purport to protect. We are operating under special licence from the Azorean government, but no one has asked the whales.

The Atlantic surges up to meet me, then sucks me in. The swell is powerful, the blue engulfing. I'm weightless and free. I duck down to avoid Drew's descending six-foot-something bulk – armed as he is with his underwater camera – and attempt to orientate myself. João told me to look up every so often towards the landmark of the vigia, but that's not so easy. Our skipper is a stern taskmaster; I feel as if he's training me on the football field. He shouts instructions at me as I paddle away, in the general direction of whales.

Through the waves that rock in front of my mask I can see, albeit intermittently, the animal's head, the plosive blow from its single nostril. The sun turns its skin grey and shiny as it bobs there, rising and falling. Maybe it's as nervous as I am. It's not easy to maintain your balance when the sea is swaying you from side to side as if you were a goldfish in a bowl being carried in a pair of unsteady hands.

Above is normality; below, everything is different. It continually surprises me, during these days with the whales, how invisible they are; like birds that vanish in mid-air, they seem to disappear in the sea. It's an impossible feat of prestidigitation. Over the waves I can see the whale, quite clearly close; under the water, nothing. Then suddenly there it is – a great big beautiful animal held in the surf, stilled within the surge as I am flailing.

To find oneself hovering over a whale's flukes, caught up in what seems to be slow motion, is truly dreamlike, because it relates to nothing that could possibly happen on land. I'm walled in by whale and water, yet at the same time entirely open to what is around me. Nothing else matters. I feel nothing bad can happen if I'm with a whale. As if its grey mass insures against all the other evils. And I feel that because I am aware, in my head, of the power of its brain as well as its body.

It's stupid to be scared in such a luminous place. Their world is bright even when ours is overcast. As the clouds slide off the volcano's slopes and into the sea, the conditions seem as uninviting as the sky, the water slate-grey, and three miles deep. But as I dive again, on the third day, it seems that a bank of lights is switched on. What appears dull from above is a floodlit field below. The sea's surface acts as a lens, both filtering and focusing the sun's rays. Under the ocean's sky, the whales' blue world is light beyond light, just as it shades into utter darkness – the profundity where they spend most of their time.

It's as if we were walking around with the night forever over us – as indeed we are, since the blackness of space is always there. The black and the blue, the dark and the light only underline the sense of scale. The whale's environment would mean death for me. But it represents life in such a vast dimension that it takes all the fear away. I might be hypnotised by these mysterious animals, persuaded to stay a little longer, just to see what happens.

Like the urge to throw myself off a cliff, the depths and their whales both appal and attract, dangerously. They have the measure of me.

Out of the obscurity, a dark shape resolves itself into a large whale, a female. From below, it is joined by its calf. As they move just beneath the surface, the young whale aims at its mother's belly to feed.

It is an intimate tableau, and I feel like an interloper, as if I were staring at a woman breast-feeding in a café. Animals as old as thirteen have been found with milk in their stomachs, the equivalent of a human teenager suckling at their mother's breast, or indeed the breast of their aunt or their mother's best friend. This is alloparental care – a shared responsibility in which even unrelated females will suckle one another's young while their mothers dive for food, as much for comfort and succour as for sustenance. These babysitters include non-reproductive or elderly females; whale tribes have a role for members which might otherwise be regarded as useless.

Such behaviour emphasises the obvious: that sperm whales live in highly developed social structures. 'Sperm whales are nomads, almost continuously on the move,' says Hal Whitehead. 'Their most stable reference points are each other.' Home to a whale is other whales. Strong social bonds define sperm whales, like elephants, which they much resemble – the one possessed of the biggest brain on land, the other in the sea; both using over-developed noses as an extension of their senses; both with ivory teeth or tusks and small knowing eyes set in wrinkled grey skin; both highly matriarchal – sperm whales might as well be elephants in the water, or elephants, whales on legs. And as elephant society is itinerant, centred around itself, so where whales are not is as important as why they are there – both to themselves, and to scientists. Hal estimates a post-hunting global population of 360,000 sperm whales spread over 316,620,000 square kilometres of ocean, 'giving a mean average density of 0.0011 whales/km^2'. Such sums cannot assess how many whales lived in the oceans before that, although, before the global spread of *Homo sapiens* over the past two millennia, sperm whales accounted for the greatest biomass of any mammal on the planet. These ancient animals might as well be updated dinosaurs, facing the same prospect of extinction.

By the time I was born, most of the world's great whales had been killed. Twentieth-century whaling devastated cetacean societies by depriving them of large males, a legacy which, given the longevity of whales, may take centuries to work out. That they manage to thrive, above and beyond all the threats they still face, is proof of the power of their natural selection and their social organisation. Given that their culture and organisation is passed on matrilineally, it is intriguing to wonder why sperm whales are so focused on the female line. One explanation is that, unlike land mammals, such groups cannot rely on males to defend them, since they may be attacked from any direction, in their three-dimensional world; and in any case, bull sperm whales travel far from the females, to remote northern or southern latitudes, in the same way that male elephants wander, returning only to breed. Thus the masculine role is reduced, and social hegemony reversed; perhaps sperm whales are truly liberated.

At the same time, however – and somewhat paradoxically – Hal and his colleagues observe that these huge clans are mostly restricted to the Pacific, perhaps for reasons of safety, since the possibility of attack from orca is greater there than in the Atlantic. The result is a social difference between the whales of these two oceans similar to that between Western and Asian humans; they are all the same species, but subject to very different cultures.

In his attempt to make sense of these shifting, clicking clans, Hal organises whale society into four categories, defined by space and number: *concentrations* spread in areas over hundreds of kilometres; *aggregations*, of ten to twenty kilometres; *groups*, over areas from hundreds to thousands of metres; and *clusters*, animals within a body length of each other. The greater of these gatherings are invisible to us, simply too big to see; we can only detect them in scientific or statistical time, as it were. But we can sense they are there, like those exosolar planets, wandering through a watery universe.

Again and again over these days at sea, as I enter the water I gradually get better at judging what the whales will do and how they might react to me. I realise how subtle the signs are, in the same way that you can see in a person's eyes what they think of you long before they might put it into words. I swim alongside a large juvenile, lingering long enough to take it all in, from head to fin, from the glowing white mandible to the chunk taken out of its caudal peduncle, a ferocious scar above its tail. Later, I see the same whale even closer, an encounter which leaves us both open-mouthed, the animal's jaw agape, slowly opening and shutting. Afterwards I wonder if it was out of stress, just as ravens will half-open their beaks in fright.

Whales would do well to fear our world. Many have marks and wounds, testaments to struggles with fishing gear or ship-strikes. 'They're tough,' says João when I climb back into the boat, babbling my description. 'They heal very well.' Their bodies appear as forbearing as their cetacean souls, although I was once rebuked for daring to presume that a whale might possess such a thing.

I think João is laughing at me.

Back down below, a young calf eyes me up, then spy-hops at the surface for a better look. Underwater, it's pale, cherubic and innocent, till it lets slip a cloud of runny poo in my face – possibly an act of defensive deception, or maybe even play. It's joined by another, equally inquisitive juvenile. Emboldened by each other, they come a little closer.

Suddenly, their number is dramatically swollen: a huge female, with an additional two calves, twirling around to appear out of the turquoise gloom. I sing to myself as I'm caught up in the crowd; I've never shared the water with so many whales. There are whales across the entirety of my vision; wall-to-wall whales wending this way and that; perpendicular, horizontal, vertical columns in the sea. More than ever, their subtle colours shine through the water; the filtered light playing on their backs, dancing on their sides. Only something so huge could be so elegant; they move more delicately because of, rather than in spite of, their mass.

Only one of these calves can belong to the large female. What I am witnessing, as these huge animals twist and turn around one another, forever touching, forever reassuring, is a cetacean crèche. For a moment it seems I might be adopted too. The mother looks at me serenely, perhaps aware of her power, while her brood, encouraged by the protection of her flanks, peer as curiously at me as I peer at them.

Then she decides it's time to move on. Gathering her charges together, she takes off into the blue, with barely perceptible acceleration. I'm left treading empty water, surrounded only by ocean.

On our final day at sea, our licence runs out. The high season is upon us, and we are no longer permitted to enter the water with the whales. We must content ourselves with seeing them from above. It is then that a group of sperm whales chooses to socialise at the surface.

It is one of the most astonishing sights I have ever seen. Slate-grey shapes in the water, they continually touch one another; it is difficult to tell where one whale starts and the other ends. Heads rise up out of the water as glossy black cylinders, bobbing with specific gravity, glistening like oil. Mouths gape with new young teeth like shiny white buds. To see them in the open air seems almost stranger than witnessing them underwater. I could be watching elephants at a waterhole, although each of these juveniles is bigger than any pachyderm.

For hour after hour they interact, heedless of our presence, adults and young alike engrossed in themselves. At one point a young whale appears to push itself over another, as if to roll over its back. Another lolls lackadaisically on its side, lulled in the motion of the waves and its playmates' micro-tides. Flukes and fins, flanks and bellies are intertwined, yielding and caressing. There is clearly sensuality, pleasure in intimate contact. I watch as two whales come head-to-head, touching brows, or more precisely, noses. How intense it must be to come so close to one's fellow whale in an almost physical transference of emotions, perhaps even ideas.

After this transcendent display, which leaves us feeling as rapt as the whales, a pair appears just off our bow. At first I think they're still playing with one another; only when they begin to swim straight at us do I realise it is a mother and calf. They continue to head for our prow, where I'm perched. I'm sure that at any moment they will shear off and dive.

But they don't. They keep coming, right up to the rib, so close that I could reach down and touch them. I can see the calf's head bobbing along, and through the water, barely beneath the surface, the mother on her side, presenting her entire length, easily dwarfing our craft. She is looking up at us, we are looking down at her. The whiteness of her jaw glows green like an iceberg through the inches of ocean that separate us. The hypnotic state of the past three hours is sundered. And so the pair pass by, and swim on.

5
The sea of serendipity

The birds may come and circle for a while…
But they soon go elsewhere. When they are
gone, the 'nothing', the 'no-body' that was
there, suddenly appears. That is Zen. It was
there all the time but the scavengers missed
it, because it was not their kind of prey.

THOMAS MERTON,
Zen and the Birds of Appetite, 1968

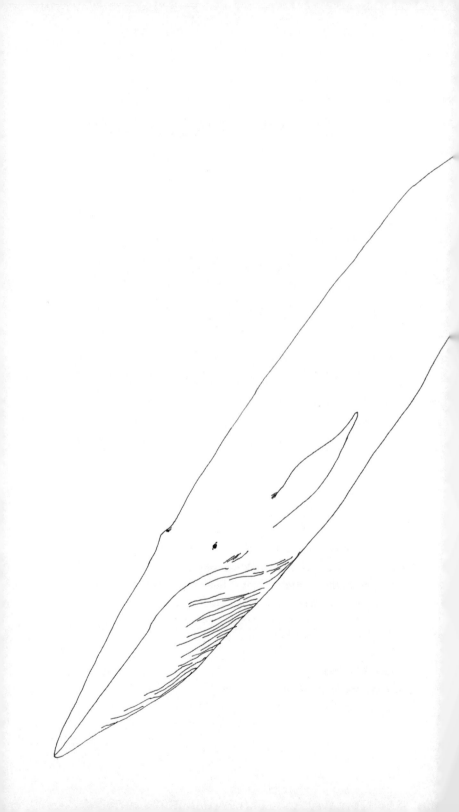

The night that fell so abruptly has yet to shift. It lies over the island like a hot, thick blanket. I awake, the ceiling fan spinning. Outside, palms are scissored out of the sky. The white noise of the sea breaks through my dreams and onto the shore.

Drew and I squeeze into the tuk-tuk, knees drawn up on the plastic-covered seat; a gilt Buddha swings from the rear-view mirror. The muezzin's call from across the street has already begun. At four in the morning people are walking to work in the middle of the road, leaving the pavement to the animals. There's a sense of a place falling apart and putting itself together. A community speeding slowly towards some determined point, in five hundred years' or five minutes' time. Bikes loom out of the dark with huge boxes of fish balanced over their back wheels; our headlight reflects the retinas of dogs stretching themselves after a night curled up in the dust.

We turn down a lane lined with low cottages in varying states of construction and collapse. The tuk-tuk jerks to a halt at a gap in the trees, out of which a man appears, gathering up his green-and-orange sarong against the rubbish that lines the footpath. Handsome, with high cheekbones, Rasika wears a peaked camouflage cap that lends him the air of a freedom fighter. We follow him past squashed plastic bottles, single flip-flops, and coconut husks. At the end of the path is a starlit sea.

Rasika steadies the *Kushan Putha* for us to board it. This does not take long, since the boat barely measures twenty feet, little more than a canoe, its fibreglass patched here and there, with uncertain stains on its once-white surface. We balance on the makeshift blue-painted planks that straddle the boat, padded with nylon-covered pillows held together by safety pins. Rasika pushes off from the beach. I feel a slight lift as the sand scrapes beneath us and the boat surrenders to the lapping waves.

As we leave Mirissa behind and move out into the bay, my eyes get used to the darkness. I see fishing boats all around, returning from their night's work, nets bulging with silvery fish. Others are racing ahead, black shapes against the even blacker water. They somehow make the dark companionable. We pass a flashing green light on the breakwater; the headland beyond is barely distinguishable. So close to the equator, day and night begin and end sharply, although the mountains delay the dawn, holding back the sun from the eastern shore. Venus flares before dipping into the blue. Up in those hills, Julia Margaret Cameron took her last look at the sky. Somewhere, too, my ancestor may have ruled over his plantation; perhaps he even sailed over these waters on his own final trip.

But all this beauty is a bit too much for me. It's still only five-thirty, so I push my roll bag into the space under the prow, climb into the cubby-hole, tug my hood over my head and curl up on the lumpy floor. It's surprisingly comfortable. Lulled by the gentle rocking, I feel the water through the thin hull as it vibrates with the outboard motor. Here in the belly of the boat, a place reserved for dead fish and smelly nets, the knowledge of what might lie below makes my berth feel warm, womblike. It's as though I were sleeping under the sea. I could easily spend the night here, covered by the omnipresent layer of salt, careless of what I might look like or how wet or dirty I might be.

None of those things matter now; they all evaporate with our intent. That subtle scrape of sand was a momentous leaving; if we carried on from here, we'd have to sail for ten thousand miles before we made landfall again, in Antarctica. This ocean links the hottest part of the earth to the coldest; in between is a vast, living arena.

My body, which had been bumping gently on the floor, rolls to a halt; Rasika has slowed the engine to a purr. Woken by the lack of motion, I slide myself out of my hole, sliding out into the light, eyes blinking. The sun has begun to break the horizon, scorching away the stars. But we have not stopped to admire the sunrise. Ahead, the surface has been broken by black shapes: the sharp fins and lithe, leaping bodies of spinner dolphin. Excitedly, I clamber to my feet.

'Come on, my boys!' I shout, as they swerve and swoop. They start to spin on cue, turning full circle in the air, showing off their sleek-striped flanks and pink-flushed bellies like salmon over a weir. Rasika starts up his engine and they're off, racing in front of our bow as their kind do the world over, as if every dolphin were performing in a perpetual oceanic Olympics.

This close, it's easy to see how they earned their binomial, *Stenella longirostris*, 'narrow long snout'. With every break of the surface their beaks poke up, prominent as broom handles, armed with dozens of sharp teeth – perfect for feeding on the fish they're rounding up below us.

Suddenly a group of common terns drop out of the sky, all angles as they dive into the same bait, whistling as if with the joy of having found such a fertile source of food. There's a series of sharper splashes – the frenetic, jagged leap of yellowfin tuna, launching themselves out of the water in their ferocious pursuit. Spiky slivers of silver with big round eyes, they look like something out of a cartoon. This concentrated spectacle will be repeated a million times throughout this ocean on this calm morning.

Then the phone rings. Rasika reaches down into a box and pulls out a 1980s-style desktop telephone, complete with push buttons and a cable trailing out the back. Drew and I look at each other in amazement; maybe the line runs all the way back to shore? Our boatman talks to a fellow fisherman, then, carefully replacing the receiver, revs the engine back into action. He navigates from a lifetime at sea, just as his fellow fishermen have come to determine the movements of their catch by the waxing and waning of the moon. Nineteenth-century whalers observed similar effects on sperm whales, which seemed to congregate around full and new moons. Such lunar assemblies may have more to do with feeding on squid, but the image of great whales guided by shafts of moon-light through blood-warm waters is too poetic to disavow. Buddhists believe any human spirit might be reborn as a bird or a fish, or a whale, or vice versa; that any animal can become another. In their cosmology, there is no God, no dominion, nor any distinction in the morality we apply to living things, since they are connected and interdepen-dent, and capable of attaining enlightenment. And in a Buddhist culture where each full moon brings a national holiday, work and play are set to natural rhythms. Inshore at dawn, men fish from wooden stilts, perched like human herons in the surf. They put out to sea in vividly-painted boats balanced with outriggers, looking more like carni-val floats. Their piratical crews dangle from the yardarms, decked out in outrageously bright sarongs as if trying to outdo one another on a Fourth of July parade. Even their nets are multicoloured – as if, under this tropical sun, nothing should be dull or monotone.

The ocean is alive. Flocks of flying fish skitter like overgrown dragonflies, dozens at a time. As we pull alongside another fishing boat, its crew yank a great silver mass from the floor: a huge sailfish, its beak a stiletto blade. One of the men pulls its dorsal fin out like a fan; it must be a metre high, electric blue and spiny. Their shiny prize took six hours to land, bucking and twisting once it was in the boat. It is a valuable catch; they won't have to work for a week. Later, we'll watch as another crew free a green turtle from their nets. Other marine wildlife are not so lucky: by-catch and overfishing are endemic problems throughout the Indian Ocean, a consequence of its crowded shores.

Further out at sea, another boat hoves into view – a mere dot on the horizon one moment, up close and larger than life the next. Sticking out from its sides like the antennae of praying mantis are huge rods made from stripped branches, sturdy enough to land a tuna. Bhangra blares from a sound system. In return for this operatic performance, the crew ask for cigarettes, employing a characteristic swivel of the head that might mean anything from acquiescence to approval – a subtle ambivalence in a country that runs on good manners. Rasika speaks volubly and rapidly in Singhalese, although it's not entirely clear whether the crew are denying that whales even exist, or whether they've seen vast schools of cetaceans, just over there, yes, there, over there; or, no, none at all.

As the land diminishes with every nautical mile, it loses all importance, reduced to a blur in the distance. To the south lies one of the world's busiest shipping lanes; on it, the same slow-moving juggernauts I see from Southampton's shore, prosaic compared to the fishermen's painted craft. At least half a dozen dark shapes dot the horizon, moving relentlessly against the sky.

Rasika points ahead, almost whispering the word we've been waiting to hear. I can't see anything. Then, there, yes, something white against the blue.

And again. It must be as tall as a house.

And again – a vertical plume in a horizontal world. A geyser erupting out of the ocean.

Kushan Putha's engine picks up speed. We're really moving now. I lean forward, white knuckles to the grabrail in front of me, head down, bent double, to an endurance test determined by the whims of a whale. Suddenly, right off our bow, the whale surfaces, sending its calling card into the air. The steamy blow is two storeys high. *Balaenoptera musculus*: the blue whale. Something so beautiful as to be unbelievable.

It is a gradual revelation. First, a dark lump, the splash-guard to the whale's blowholes. Then, rolling behind its wedge-shaped head, its body – no more a mere body than a skyscraper is a building; a living construction that leaves words in its wake.

I've only seen this creature as a shape against a white page, laid out with the rest of its relatives in the austere context of a field guide, or modelled in a museum. The reality is so far removed, so enormous, that the only thing

I can fix on is the colour. It really is blue, at least in certain lights: a deep, petrolly blue, or the blue of a swallow's back; the concentrated essence of water, with an added iridescence. Like sperm whales, these chameleons seem to change colour with every angle, every glimpse. One moment they're mottled grey and turquoise, organically patterned like lichen on a beech tree or the sheen on a plate of zinc; the next, the blue becomes so black it is as deep as the lacquer on a Japanese chair. This illusive quality makes the whale even harder to comprehend – as if that weren't difficult enough already.

Only afterwards, looking at my photographs, do I recall what I must have seen but not registered in the moment: that with its forward motion the animal drags the sea down with it, as if the water were parting to make way for the whale. Its imperial progress demands nothing less. It is animate, yet not set apart from the ocean; it is the ocean itself. I think of the visible rays in medieval paintings that connect God's gaze to his saints, as Drew tells me about a scientist who has devised a laser that, by bouncing off either end of a whale, instantly gauges its size. All I have are my eyes.

Then, as the sequence plays out, something almost ridiculous: the animal's dorsal fin, risibly small, an afterthought of sorts. The stubby bump only serves to make what went before appear even bigger. It is followed, inexorably, by the thick tailstock – the caudal peduncle – restoring to proper proportion an animal the size of an airliner. Densely muscled, it is charged, ready to to rise high into the air and jack-knife down below.

But first we must wait as the whale blows and rolls through the water. Everything is stilled by expectation, diminished by this deferred miracle. All around the world people are going about their business; we are watching a blue whale about to dive.

Subtly, the whale changes momentum, goes up a gear. You can't put your finger on the moment that it does, but it's obvious when it starts to happen. You feel the animal flex and ripple, drawing up and pulling down.

Suddenly its flukes are held sharp against the sky, as though a giant sail had been raised out of the waves. They are utterly enormous, broad and of a piece, yet at precise right angles to the animal, like a plane's tail fins. They hang there for a second, wobbling with their magnitude and the power of the body to which they are connected. They announce the end of the encounter. With this final flourish, our audience is over, leaving us wanting more.

There's something sexual about whales, Drew and I agree. Sleek, sensual and untouchable, they are the ultimate tease. It is that which draws you on.

They may be the biggest animals that ever existed, but the blue whales of Sri Lanka have no place in the island's culture. Roman maps depict a promontory on its south-east coast named *Cetcum Promotorium*, the Cape of Whales; Pliny, in his *Natural History*, written in the first century AD, described Ceylon as 'banished by nature outside the world', and therefore free of the vices of other countries. But beyond the influence of the West, the whales lived on in unhunted innocence. This island was a kind of Eden – it was also said to be the place where the whale spat out Jonah – just as whales were monsters at the edge of the map, islands in themselves in medieval myth, gathering roots, shrubs and trees on their backs as they aged: 'As if the greatest mass of sea-weed lay/Beside the shore, with sand-banks all around.'

It took a thousand years or more for the blue whales to resurface in these waters, in human sight, at least. My brother-in-law, Sampath, first told me about the whales which seemed to have miraculously appeared with the ending of the civil war and the lifting of military restrictions on the coastal waters in 2009. Until then the whale was regarded, if at all, as a stealer of fish or a destroyer of boats. Now they have become a sight to see, rather than fear, although later, when I asked a school assembly in Galle if they'd ever seen a whale, only half a dozen hands went up, despite the fact that such stupendous creatures swim only a few miles off the beaches on which the children play.

That day, our first at sea, one blow was joined by another, then another, and another. We rushed this way and that, as whale after whale appeared in every direction – two dozen, maybe more. Their plumes shot up all around us like watery fireworks. Not far beneath the surface, the whales were busy scooping up the millions of krill they need to eat each day. I could even smell it, a definable change in the air – the characteristic scent of dimethyl sulphide gas given off by the phytoplankton on which their prey feed – and with it, the possibility of whales.

All around us were mothers and calves, the young coming close to see what we were. No one has yet identified where these whales go to mate and calve, but Asha de Vos, a young marine biologist who has been studying this population, believes it is close by. She speaks passionately for the animals, which, as a Sri Lankan, she sees as her responsibility. They have their own specific, *Balaenoptera musculus indica*, and may even be a subspecies of their own, a resident, rather than a migratory population, which has developed to exploit this fertile territory. With the continental shelf so close to the tip of the island at Dondra Head, the upwellings created by the meeting of deep cold and shallow warm water provide the perfect environment.

'He is seldom seen,' Ishmael wrote of the blue whale, 'at least I have never seen him except in the remote southern seas, and then always at too great a distance to study his countenance. He is never chased; he would run away with rope-walks of line. Prodigies are told of him. Adieu, Sulphur Bottom!' he said, employing the nickname that referred to the parasites which often coat its belly like mustard-yellow paint. 'I can say nothing more that is true of ye...'

The display board in the Natural History Museum in London, in the shadow of its own illusory blue whale, seeks to prove Ishmael wrong. It tells the story of the twentieth-century cull: from almost nothing to tens of thousands in 1939 – the year in which the museum's model was first displayed – when forty thousand were taken in the Southern Ocean.

Until then they'd been too fast for humans. That was why the colonisers of Sri Lanka – the Dutch, the Portuguese, the English, all whaling nations – did not even try to hunt these whales. It was another reason for their absence in the island's history: their sheer size and speed eluded human contact, as if they were too fast and too big to see. For a Buddhist, the sense of being there and not being there would have been perfectly understandable; for others, the distance was a challenge to be closed, by explosive harpoons fired from high prows. New factory fleets from Norway and Japan and the Soviet Union did their work. By the mid-twentieth century, the hunt was at its height; hundreds of thousands of hunted whales were under-reported by the Russians. Under Stalin's Five Year Plans, each harvest had to be better than the last, and whales became part of that relentlessly expanding demand.

Great fleets of whale-catchers scythed through the seas, one of them, the *Storm*, captained by a woman, Valentina Yakovlevna Orlikova. Her photograph, published on the cover of a 1943 magazine, *Soviet Russia Today*, shows Orlikova with high cheekbones, dark hair and svelte figure in uniform, gold braid on her cuffs and a collar and tie at her neck. An even more glamorous photograph of her appeared in *Harper's Bazaar*, causing Anaïs Nin to fall in love with this 'Hero of Social Labour', a Soviet Ahabess armed with a missile-harpoon. In forty years' hunting in the Antarctic, Orlikova's superiors reported a total of 185,778 blue, fin, sperm, right, grey and other whales; the true figure was 338,336. In a cold-war world of misinformation and opposing ideologies, whales were the ultimate losers.

In 1963 the first 'International Symposium on Cetacean Research' was held in Washington, DC, attended by scientists from many different disciplines. The Symposium sought to gather the latest knowledge about whales, and present the case for the fragility of their species. It marked a new awareness and urgency; its chairman, L. Harrison Matthews, appealed for 'some friend of science' to supply the means to fit out a vessel of discovery, to fund a search for answers to these questions before it was too late. 'The cost of one long-range missile would cover the whole project,' he added. It was a telling comment, one year after the Cuban crisis and its threat of nuclear war. By the time Remington Kellogg, director of the Smithsonian, came to unveil his museum's own life-size blue whale in 1969, he could deliver to the assembled audience an updated toll, accurate to that precise moment –

329,946 blue whales dead

– most of them on his own watch, as the half-life of whales ticked away to doomsday. It was only now, now that it was too late, that this well-meaning scientist and instigator of the International Whaling Commission, who had devoted his career to collecting data, realised that he had spent most of his life 'working for the enemy', as D. Graham Burnett, the historian of science, reflects. It was a neat, horrific formula for hip-booted scientists standing knee-deep in the gore of whaling stations as they traded whisky for foetuses:

Collecting data while animals died + animals dying so that they could collect data = science as self-fulfilling prophecy.

Even now, in the bright new era of modern cetacean science, we insist on assimilating the whale. We may have instituted less violent means of investigating their physiology and movements, yet we endeavour to dart or tag them, tracking them by satellite with GPS to provide us with neat data sets although these techniques can be deleterious to the animals themselves, causing infection or even restricting movement. In effect they become remote-controlled models to our will, mapped, suborned and co-opted into our world, rather than left alone in theirs.

Out in the Indian Ocean, we watch as a huge vessel looms into view. The grey-painted, block-like ship with shuttered windows was previously used to transport troops during the war. Now it has been commissioned to take passengers out to see the whales. 'Big sound, very bad – whale diving,' Rasika observes pithily, as the navy boat steams ahead. Later, we learn that one of the passengers on board its first voyage was Vladimir Putin, who some months before had been photographed in the Barents Sea, darting a whale with a crossbow in order to tag it.

No sooner have I plunged into the water than the whale has gone. I hang there, alone, suddenly aware of how vulnerable I am, open to anything that might approach, from any direction. My rising fear is hardly calmed by the knowledge that the ocean's bottom lies a mile beneath me.

As I look down through the windscreen of my mask, I see a snake-like animal curling and writhing. Its spine seems luminous, repellent, as it twists about in the water column. Powering with my fins, I return to the side of the boat, turning my back to it as I bob just below the surface, as if its flimsy hull might shelter me from the unknown. After a few more minutes scanning the darkness below, I'm glad to give up and haul myself back into the *Kushan Putha*.

The ancient name for Sri Lanka is Serendip, from which came our serendipity, coined in 1754 by Horace Walpole after reading a translation of the sixteenth-century Italian fairy tale *The Three Princes of Serendip*, whose characters 'were always making discoveries, by accidents and sagacity, of things they were not in quest of'.

'What is your name? Where do you come from?' say the boys on the beach. After a while, I invent my own story. I'm married, with five children; I'm a French soldier; I don't understand. Men here stand with their arms around each other, almost more beautiful than the women; everything is about display. Giant blue-and-red kingfishers skim over the river. Peacocks perch in trees like outrageous, overgrown pigeons.

At Weligama, a tiny island lies a few hundred feet offshore, little more than a palm-topped rock. To reach it, there's no need to take a boat: the water that separates it from the land is seldom more than a few feet deep. Clustered on its summit is an eccentrically shaped villa, barely visible through the trees. Invited there one evening, we waded through the warm water, flaming torches spiked in the sand to guide us across. Reaching the white-painted jetty on the far side, we passed up a ferny path and into the house. Off an octagonal hall lay eight doors; each room might have opened on eight others. Around the whole building ran a wide verandah. Entertainment appeared to be the entire function of the place; it was a fantastical confection, not unlike Walpole's own gothic villa, Strawberry Hill, on the banks of the Thames.

The house was built in 1927 by the self-styled Count de Mauny-Talvande. Son of a banker, rather than an aristocrat, he was born Maurice Maria Talvande in 1866 and was educated by Jesuits at St Mary's College in Canterbury, Kent, where he met George Byng, the son of the Earl of Strafford. In 1898, Maurice married George's sister, Lady Mary. It was a splendid occasion, attended by the Princess of Wales, to whom Mary was lady-in-waiting, along with other European royalty whose houses were to fall in the coming decades.

The marriage was not a happy one. Having established what he called a university in a château on the Loire in France, teaching the sons of the British aristocracy, Maurice was accused of making advances to Oliver Brett, the son of Viscount Esher. His hasty departure was seen as an indication of his guilt; it was said that his wife would send out remittances to her estranged husband, effectively paying him to stay away. Lured to Ceylon by the sight of a flame lily growing in a Bournemouth garden, de Mauny looked for 'the one spot which, by its sublime beauty, would fulfil my dreams and hold me there for life'. In 1925, at Weligama, he found it: 'a red granite rock, covered with palms and jungle shrub, rising from the Indian Ocean – an emerald in a setting of pink coral'.

Having bought the island for two hundred and fifty rupees, de Mauny named it after the Greek name for Ceylon, Taprobane, meaning 'garden of delights'. It was a suitable retreat for a self-invented man. But it also had a darker reputation, as a dumping ground for cobras, put there by Buddhist monks forbidden from killing them. If Sri Lanka was the original Eden, then Taprobane was its estranged scion.

The house de Mauny built on this two-and-a-half acre site was extraordinary, even for this extraordinary place. Hanging gardens tipped over its terraces, surrounding the octagonal villa which acted as a kind of compass in this improbable coordinate. To the north was Ceylon, to the east the Bay of Bengal, to the west Arabia and Africa. To the south was the Indian Ocean, and beyond, the South Pole. All the world might be seen from such a place, while its prying eyes were excluded. The setting only encouraged de Mauny in his ambition. He sought to evoke a grand tour, as Christopher Ondaatje wrote, with 'echoes of the Italian lakes, the isle of Capri, some details of the Kandyan style, Alhambra styled carved wooden pillars and even some elements of the Vatican gardens'.

At the heart of this exquisite palace lay its Hall of the Lotus, open to the skies and lined with panels of blue and gold. It was entered through a theatrical arch dressed with blue silk curtains in the *art nouveau* style (there were no doors); its dome rested on eight sky-blue pillars. The entire house resembled a magical lantern set down on the tiny island, illuminated by golden light pouring through amber-coloured glass blinds. Even the iron gates that admitted the chosen to this artificial paradise were surmounted with brass-headed peacocks through whose turquoise eyes, it was said, one could see the vast Indian Ocean sprawling through time.

To fill his house de Mauny designed his own furniture, made in Colombo to a French style, some pieces of which survive even now, despite floods and hurricanes. Work proceeded slowly, and it was not until 1931 that de Mauny took up residency in his lavish abode. Throughout the thirties, European nobility called on Taprobane – the same society that had rejected the would-be aristocrat back home was lured to his island, as if what was taboo in Western circles could be allowed far from its salons and drawing rooms. But de Mauny had only a few years to enjoy his grand entertainments; he died of a heart attack in 1941, and with his death his vision seemed to slip back into the sea. A year later his son, a naval commander who lived in Hampshire, sold the island.

————

One afternoon in 1949, in his country home near Salisbury, Wiltshire, where his family had occupied the great mansion at Wilton since Shakespeare performed his plays there, David Herbert, second son of the Earl of Pembroke, was showing his family albums to his friend, the American writer Paul Bowles. Turning the pages, Bowles happened on a photograph of Taprobane, taken when Herbert had visited Ceylon in the 1930s.

Bowles was transfixed. He saw the 'tiny dome-shaped island with a strange looking house at its top, and, spread out along its flanks, terraces that lost themselves in the shade of giant trees'. Bowles was an inveterate wanderer in exotic places; he was also the newly successful author of *The Sheltering Sky*, a nightmarish story set in Morocco, where he and his wife Jane were at the centre of a group of bohemian expatriates. But Bowles, who was naturally reclusive, had begun to find their company claustrophobic, and seeking to escape, determined to go to Ceylon. He sailed from Antwerp that December, on the freighter *General Walter*. As the ship drew nearer to the island, he found himself recalling Kafka: 'From a certain point onward there is no longer any turning back. That is the point that must be reached.'

Bowles was referring to the book he was trying to write, although he might have been addressing his psychological state. When he finally saw Taprobane, the black-and-white photograph he had glimpsed in an English stately home became reality, and it exceeded his expectations, 'an embodiment of the innumerable fantasies and daydreams that had flitted through my mind since childhood'. Two years later, Bowles bought the island for five hundred dollars. It became his version of Cuthbert's Inner Farne, albeit rather more worldly and sensual.

Bowles relished his retreat. 'The maid polished the furniture and filled bowls with orchids. The gardener fetched things from the market in the village on the mainland. Another man, a Hindu, came twice a day to empty the latrines, as there was no running water on the island. Life moved like clockwork,' he claimed, 'there were no complications.' In the evening, Bowles and his lover, Mohammed Temsamany, would wade across the water in their bathing suits, their servants carrying their masters' clothes in bundles on their heads. On the other side, they would dress and set off in the direction of the devil dancers whose displays had come to obsess the writer. To him their convulsions seemed a kind of shock therapy; the masked figures in the darkness, lit by their flaming torches, driving out 'demons of pain, psychosis and bad luck by inducing such terror in the subject that he will automatically expel them'.

At night Bowles lay in bed, listening to the waves crashing on the island's seaward cliffs, contrasting them with the softer sound they made on the sandy bay. It was the greatest luxury he could imagine. He was untroubled by the sharks that swam around the reef, or by the enormous turtle that haunted the rocks, rising to the surface occasionally like 'a floating boulder', its great domed back an indication of its age, so the gardener said. During the hours of darkness, flying foxes hung in the trees, to be chased away at daybreak by flocks of crows. 'Once they had done that and remarked about it with each other for a while, they flew back to the mainland. But the bats never returned before dark.'

Sitting at one of the Count's desks, cigarette holder in his mouth, fingers poised over his typewriter, Bowles wrote his novel *The Spider's House*. It was set in Morocco, but its title might have applied to his current refuge. His wife Jane was much less enamoured of the island. She took to drinking, hard, and particularly objected to the flying foxes – they seemed to be leathery demons to her – and couldn't wait to leave. Perhaps she sensed a place haunted by its past. The dark island of Ceylon itself may have been where all the devils were. As a young boy growing up near Colombo, my brother-in-law Sam saw a burning light at the end of the garden one night. As he watched, the apparition zoomed closer and closer towards him, taking on the shape of a human head engulfed in flames.

With the arrival of the high seas of the monsoons, and the high taxes levied on foreigners, Paul Bowles was also driven out. According to one writer, Richard Hill, he became unpopular with the locals, not least on account of his use of hashish; at night they came to whistle and throw stones at the island and its unwelcome tenant. After only two years on Taprobane, Bowles sold up and retreated to Tangier, from where I received polite replies to my enquiries, neatly typed on cigarette-paper-thin sheets, as well as an invitation to visit which I never took up.

Nor did I see any devils on Taprobane, although for all I knew there might have been hundreds hiding in the lush vegetation. Guests drifted under the thirty-foot-high cupola of the Hall of the Lotus from which the villa's rooms radiated like the hands of a clock, and where any manner of other lives might be being lived out. De Mauny's elegant furniture still stood against the walls. In the darkness, I swam around the rocks, the light from the torches flaring on the lapping waves. Only when it was time to leave did we realise how far the tide had risen. Wading across the water, which was now above thigh-height, I had to support an older lady, elegantly clad in an immaculate gold-embroidered sari, as she picked her way through the swirling cocoa-coloured sea like a water-logged bird-of-paradise. I wouldn't have been surprised, looking back over my shoulder, if the entire island had disappeared behind us, sinking into the Indian Ocean.

In an interview for the BBC conducted at the New York World's Fair in 1964, the writer Arthur C. Clarke spoke, in his heavy Somerset accent, of things to come. 'The only thing we can say of the future is that it will be absolutely fantastic,' he declared. He went on to make a specific prediction for the year 2000, even though he thought 'it may not exist at all'. Not through nuclear war or a 'new stone age', he said, but because of incredible revolutions in communication.

Clarke foresaw a world 'in which we could be in instant contact with each other, wherever we may be...

for a man to conduct his business from Tahiti just as well as London, independent of distance'. By then the globe would have 'shrunk to a point where men will no longer commute, but communicate', and the modern notion of the city would have been abolished – although at the same time Clarke feared that the world would become one great suburb.

The prophetic air of the black-and-white film – faintly undermined by its very British voiceover – is animated by a glimpse of a working model of that future world, one in which 'from the heart of what was once tropical jungle will spring new and glossy cities', with superhighways and laser-felled trees, settlements at the poles, and under-water hotels and cars; an entirely inhabited world, the vast suburb that the writer dreaded.

By then Clarke himself had retreated from the modern world to live in Ceylon, ostensibly to pursue his passion for scuba-diving, although he may have been drawn by its other attractions. Inspired by his love of the sea, he also conjured up a more disturbing vision: 'In the world of the future, we will not be the only intel-ligent creatures. One of the coming techniques will be what we might call bioengineering – the development of intelligent and useful servants among the other ani-mals on this planet, particularly the great apes and, in the oceans, the dolphins and whales.' Clarke thought it a scandal that man had neglected to domesticate any animals since the Stone Age. 'With our present knowl-edge of animal psychology and genetics, we could cer-tainly solve the servant problem,' he said, although he also foresaw animal trades unions, 'and we'd be right back where we started'.

It's difficult to tell how old Anoma is. He may be half my age, or older than me; he has a youthfulness about him. Since he was a boy he has trained himself to be open to the natural world. He was born here in Galle, knows it intimately. Walking through the remains of its Dutch ramparts, he points out a giant ficus whose roots seem more solid than the massive stone walls of the seventeenth-century fort, fingering its way in clusters to the cracks, slowly pulling the place apart.

We emerge out of the deep, gloomy gateway that regulates those seeking admission to Galle's ancient precincts, and which might well have been built to allow the passage of an elephant. Above us the museum rises like a cliff face. It is punctuated with arched windows, and seems impossibly old, ark-like; indeed, within its dark belly lay the skeleton of a Bryde's whale, articulated and displayed there till the sea came to reclaim it. Anoma shows me the level to which the water rose that day, a memory marked in feet, although every inch measured out disaster. The slow withdrawal of the water. The fish flapping in the pools. The silence. Then, roaring out of the ocean, an unbelievable rising, an incomprehensible volume produced by one tectonic plate sliding under another. Below, whales had already fled the submarine sound waves; above, millions of lives were about to change.

That morning, 26 December 2004, Anoma was sitting in his office in the Lighthouse Hotel, watching the sea swell over a great rock called Whale Head. Now it was lost in the surging waves that engulfed the lower floors of the luxury hotel. Two of Anoma's colleagues, on their way to work, were simply washed away. Bicycles, tuk-tuks, animals, cars, houses, boats, stupas, people; there was no discrimination in the natural destruction, nor any dominion, either. Nor would it have occurred to anyone to worry about the inhabitants of the sea as those terrible tremors coursed through its depths.

In a recent study it was found that, after an earthquake off California measuring 5.5 on the Richter scale, a single fin whale covered thirteen kilometres in twenty-six minutes as it attempted to flee the source of the two-hundred-decibel sound. In Sri Lanka, animals on land were said to have left the coast for higher land long before humans had warning of the tsunami. At Taprobane, the flying foxes left their caves in broad daylight, of their own accord. Later, when new shocks threatened the same waters, naturalists at sea, unaware of earthquake alerts, were mystified to see every cetacean, from blue whales to spinner dolphins, disappear. Meanwhile, the water in hotel swimming pools lurched and lapped like little oceans.

On the road back to Colombo, posters and plastic bunting proclaim rival allegiances; political symbols are spray-painted onto the tarmac. On the city's outskirts at night, young men march behind wire fences. 'People are afraid,' my friend tells me. Suddenly it seems even darker outside. What do they do, those slim young men? What will they be expected to do?

And out at sea, the whales leave black holes behind, taking everything in their wake.

6
The southern sea

We cannot think of a time that is oceanless
Or of an ocean not littered with wastage
Or of a future that is not liable
Like the past, to have no destination

T.S. ELIOT
'The Dry Salvages', *Four Quartets,* 1941

On 29 January 1831, the convict ship *John I* arrived in Van Diemen's Land. On board was a new consignment of two hundred prisoners from Britain. Among them was a twenty-nine-year-old cabinet maker from Newcastle-under-Lyme, described as a 'very bad old offender'. Two years previously he had been convicted of stealing fowl, for which he was given a month in prison. In 1830 he repeated the offence in Stoke-on-Trent, this time with nineteen hens and one cock, and was tried for his crime at Stafford Assizes. He was sentenced to seven years' transportation. His name was James Nind, and he was my distant cousin. Like my other relatives of that same name, he had travelled far from England, but unlike them, he had no choice in the matter.

I know little of James's life before his conviction – only that he came from the extended family of Ninds who had flourished in middle England – but his crime has bequeathed him a kind of immortality in the penal records of Britain and Tasmania. After his conviction, he was kept in gaol with his fellow transportees until they were taken south, chained hand and foot, to Portsmouth. There they were decanted into prison hulks – slowly rotting, superannuated Napoleonic ships named *York* and *Leviathan* which lay stranded in the mud as if serving their own sentence, overbuilt arks with shed-like structures and redundant rigging from which laundry fluttered in the breeze; in Dickens' *Great Expectations*, Magwitch escapes from such a hulk before being transported.

Once on board, the men were sent deep below. At night, after lockdown, they were left to their own devices. By day they were put to work in Portsmouth's naval dockyard. Every kind of vice ran riot in such circumstances, and although James Nind's behaviour on his hulk was reported to have been 'good', he was probably relieved when the time came to embark on his voyage for the Antipodes.

John I sailed from Spithead on 9 October 1830. Along with its cargo of convicts, it carried a detachment of thirty officers and men of the 17th Regiment, as well as eight women and nine children; the journey took 106 days, with up to five prisoners sharing one sleeping berth. Most had been found guilty of theft, although five men had been convicted of rape, and were to serve life sentences. In some cases, transportation allowed magistrates to impose more lenient punishment on those they might have been forced to send to the gallows for such crimes as James's. Yet the voyage was a trial in itself, for innocent and guilty alike. Although the log kept by the *John I*'s surgeon maintains that prisoners were allowed on deck during fine weather, had their manacles removed if they were suffering from ulcers, and were given lime juice to prevent scurvy, their true situation was appalling. Men were locked together and treated little better than slaves. In bad weather and when disease ran rife below decks, the journey took on the aspect of a nightmare; the surgeon serving on the *John I*'s previous trip had thrown himself overboard 'in a fit of lunacy'.

After more than three months at sea spent mostly in the dark, the prisoners emerged, blinking in the slanting southern light of Van Diemen's Land. Their arrival coincided with the stern governorship of Sir George Arthur, a man who believed that surveillance was the key to control. He presided over an island which was, in effect, an open-air Panopticon – the all-seeing penitentiary proposed by the radical philosopher Jeremy Bentham, who regarded prison as 'a mill for grinding rogues honest'.

After the establishment of Botany Bay as a penal colony, Van Diemen's Land had become a last resort for reoffending criminals, sent to Macquarie Harbour, a 'Place of Ultra Banishment and Punishment' set in a remote inlet on the far side of the island. Conditions there were so appalling that rather than endure its tortures, men would commit murder in order to be hanged in Hobart. Others were held in complete isolation. Sent to the end of the world only to be kept in utter silence and complete darkness, some simply lost their minds.

But like Port Arthur, the infamous prison on the Tasman Peninsula, such horrors were reserved for recidivists. James's servitude in Hobart might have been quite endurable. Convicts were employed in public construction works or as bonded labour, and although theirs was a state only a step above slavery, many were well treated. James was assigned to Captain John Montagu, 'on loan', and worked as a carpenter, biding his time. Men serving seven-year sentences such as his could apply for a ticket-of-leave after four years' good behaviour.

Unfortunately, my cousin did not behave himself. On the morning of 7 November 1831 he was found drunk in a public place, for which he was sentenced to twenty-five lashes – a punishment that could tear the flesh from a man's back. Luckily for James, the sentence was suspended (possibly because Montagu was also justice of the peace). Four years later, on 27 July 1835, he was found guilty of another misdemeanour, solemnly inscribed in the 'Black Book' of convicts' records: 'Indecent conduct in a Public Street and exposing his Person while Making Water.'

James's latest indiscretion earned him a suspension of his ticket-of-leave, which had been awarded in honour of the Queen's birthday that year. But he did not learn from his foolish ways. On 25 July 1836, a third infraction is recorded under his name: 'Misconduct in being in an uninhabited house with a female prisoner after hours.' James had the luck of the devil, or perhaps a certain charm. For this assignation he was jailed for fourteen days, his ticket-of-leave once more suspended.

The female prisoner was Sarah Worth, an auburn-haired, freckle-faced young woman from Cheshire who, along with her younger sister Mary, had been transported in 1831 for stealing clothes. In November 1836, his freedom restored, James was given permission to marry Sarah. They lived in Cascade Street, Hobart, and went on to have one child, Ellen. Sarah died in Hobart in 1852, but James lived to the great age of ninety, having moved to Corowa, New South Wales, where he died, never having returned to England, in 1891.

Orion wheels high overhead, but here his sword points upwards. This is a place of beauty and unease. I'm further south than I have ever been.

Hobart is a frontier town, swept by polar winds. Down at the quayside, bright-orange icebreakers arrive, bringing back scientists dazed by their first glimpse of anything other than whiteness for half a year.

I leave my room in a former whaling captain's house and walk out onto a dark street just before dawn. At the end is a circular close, lined with low cottages. It might be an ordinary suburb, if not for the fact that these were also homes for whalers, built by convicts from convict bricks. As the sun rises, the air is so clear and unfiltered by pollution that the distant mountain seems to speed towards me. The empty streets turn back time.

I swim from an urban cove, reluctant to push out into water once inhabited by whales – the southern rights or black whales, massive, ponderous cetaceans that came up the Derwent River to their age-old mating and calving grounds. I've watched these enormous animals, or at least their northern cousins, rolling around one another within sight of the Cape Cod shore, so engaged in their sensual play as to be ignorant of the human spying on them. Such self-involvement made it easier for their ancestors to be slaughtered in their thousands. Whaling was the commercial foundation of the Antipodes, closely allied to the same process that saw its islands become a convenient dumping ground for Britain's moral rejects.

In 1804, David Collins – a naval officer who had been sent to settle a new penal colony on Van Diemen's Land, having been fervently lobbied by Bentham to put his Panopticon into practice there – had written to Joseph Banks from Hobart, telling him that the river was full of whales, so many that three or four ships could have filled their holds with oil. Later that year the adventurer Jorgen Jorgensen was sailing as first mate on the English whale-ship *Alexander* when he claimed to be the first to kill a whale in the Derwent.

Had its brothers and sisters been warned by the violent death to which their near relation was thus subjected, and avoided the fatal spot for the future, I would have little hope of living in the grateful remembrance of future whalers; but the contrary is the case, for the destruction of one apparently attracted many hundreds of others to crowd up and incur the same fate, and the rising city of Hobart Town is yearly and rapidly becoming enriched on their oleaginous remains.

The estuary was so full of whales that it was dangerous to navigate the river, and the governor complained of being kept awake by their lusty blows. They would soon be silenced. By the 1840s, as a result of thirty-five whaling stations around the bay, there were only a few animals left for them to kill. Hobart's hunters turned to deeper waters, and the sperm whales which swam in them.

Yet no one had come here for the whales; at least, not at first. Two hundred years before, in 1642, Anthony van Diemen, Governor-General of the Dutch East India Company, sent Abel Janszoon Tasman to map a place hitherto named simply Beach. In a remarkable omission, Tasman missed Australia altogether, and only brushed against its largest island, where he attempted to land at what would become known as Adventure Bay, but was driven back by storms. Having succeeded in planting the Dutch flag on North Bay, he sailed on to New Zealand.

It took more than a century for another European to arrive. In March 1773 Tobias Furneaux's ship, *Adventure*, separated from its sister ship, James Cook's *Endeavour*, sheltered in the bay to be named after it. At that point, no one knew that Van Diemen's Land was an island, nor did Cook himself land there until his third and final voyage, when he carved his name on a tree: *Cook, 26 jan 1777.*

Adventure Bay lies on Bruny Island, which hangs off the coast, into the Tasman Sea, due south of Hobart. Turquoise waves break on its beach, having crossed the Pacific to get there. On an outcrop of rock stand two huge blue-grey eucalypts; they were here when Cook pulled ashore. Despite its apparent civilisation, this coast is largely untouched: those journeys of discovery might still be in progress today, sailing up and down the coast.

At the end of the bay, I walk through an out-of-season holiday park of cabins. One large hut, containing the 'facilities', is bedecked with whale ribs and vertebrae. The path takes me through a closely-packed stand of eucalypts, swaying over my head. A deciduous tree's leaves are angled to make the most of the light; a eucalypt's hang vertically, as if they'd had enough. Out of the shadows and onto the beach, a white wallaby appears, staring at me with albino eyes. Along the shore, a desultory pile of rocks testifies to a former whaling station. At the tip of the beach is a turf-topped rock, Penguin Island. Here Cook took his leave of the land, his last step on Australian soil.

It was for such wonders and desolations that men chose to sail around the world. When Cook engaged Joseph Banks – or, more accurately, agreed to his joining the expedition as its underwriter – the influential and wealthy young botanist declared: 'Every blockhead does the Grand Tour. My tour shall be one around the globe.' These voyages too were extensions of Georgian culture and taste: for Cook's second voyage, Banks proposed that they should be accompanied by the portrait painter Johann Zoffany, as if it were a society outing. And although Zoffany – famous for his staged 'conversation pieces' – did not make it to the Pacific, he nonetheless produced a heroic, neo-classical image of the story's finale, depicting the slaughter of Cook on a Hawaiian beach in the style of a Grecian frieze. Soon after, a Parisian manufacturer produced 'Captain Cook Wallpaper', ready to paste on one's drawing-room wall.

More than a thousand years before, St Augustine had queried the practicality of travelling to the other side of the world – 'it is too absurd to say, that some men might have taken ship and traversed the whole wide ocean' – or that men could even live in a place in which the word of God had not been heard. Now the Age of Enlightenment had been alerted to the reality of an upside-down world. To Cook and Banks, the strange new species that they encountered there presented a challenge of identification and order. To those who followed, they were ready for the taking.

The Tasman Peninsula is a great mass of Jurassic rock, the highest sea cliffs in the Southern Hemisphere, a spectacular wall of black dolerite. Somewhere behind them lies Port Arthur and its prison ruins. Tim, our twenty-two-year-old skipper, clad in board shorts and boots, brings

the nine-hundred-horsepower rib to an abrupt halt and hands the controls to his first mate Ben, who is even younger than his captain.

'This is awesome,' Tim enthuses, as we look up at the ancient strata and crevices. Everything is open and sharp here, as if the world were still being built. Fur seals loll on every ledge, slipping into the water to arc in and out of the waves. Their name, *Arctocephalus pusillus*, 'bear-headed little one', perpetuates an error, having been first described from an illustration of a pup. In truth they're the largest of all fur seals, and seem to revel in their status, with their upturned noses and haughty profiles – for all that they are forever picking fights with one another, flashing their eyes and baring their teeth. Were I foolish enough to approach one, I'd soon discover that a fur seal can move fast, even out of the water. Faced with such a situation – so the pages of my field guide tell me – I should maintain eye contact and slowly back off, since bull seals see an upright human as a threat.

Perhaps they have good memories. In Tasman's and Cook's wake came the sealers, supplying the Western world with fur. The elegant ladies and gentlemen who preened themselves on Piccadilly or the Unter den Linden had no idea of the desperation that lay in the sleek pelages which protected their necks on a chilly winter's afternoon. Nor could they know the nature of the men who peeled the pelts from their still-living owners: lawless men, many of them escaped convicts who kidnapped Aborigine women, keeping them captive as sex slaves on places such as Kangaroo Island.

Sealing began in Australia in 1798. Such was its ferocity that within thirty years, three species – the New Zealand fur seal, the Australian sea lion and the Southern elephant seal – were all but extinct. Three-quarters of a million animals died, many skinned alive. Only the Australian fur seal remains in Tasmania. As we sail around the cliffs, ledge after ledge is filled with these restive animals, lying belly to belly, disputing the perfect spot. Tim steers the rib into the fractured, barnacled shore. Peering down into the clear water, in the opening and closing gap between rock and boat, I can see huge stems of swaying brown kelp, twenty metres tall, a playground for the seals that anchor themselves to the stems with their surprisingly dexterous hind limbs. We drift dreamily over the thick fronds as though over a gelatinous forest.

A Shy albatross soars out of the sky – the first I've ever seen. No other bird could be so shaped to the sea, 'a regal, feathery thing of unspotted whiteness, and with a hooked, Roman bill sublime', as Melville wrote when he saw his own first albatross, flying on 'vast archangel wings as if to embrace some holy ark'. Like any other reader after *The Rime of the Ancient Mariner*, Melville was alerted to the bird's romance by Coleridge's poem. But the poet (who called himself a 'library-cormorant') drew his story from George Shelvocke's *A Voyage Round the World by Way of the Great South Sea*, published in 1726.

Most sailors regarded the sight of an albatross as auspicious. As it ages, the wandering albatross becomes almost entirely white, not unlike the Risso's dolphin or the beluga whale: a ghostly soul-bird, invested with the spirits of departed sailors or the whiteness that appalled Ishmael. Shelvocke's second captain, Simon Hatley, saw the black albatross as an evil sign, 'observing, in one of his melancholy fits, that this bird was always hovering near us, imagin'd, from his colour, that it might be some ill omen'. As Hatley took aim –

> 'God save thee, ancient Mariner!
> From the fiends, that plague thee thus! –
> Why look'st thou so?' – With my cross-bow
> I shot the ALBATROSS.

– his commander looked on in a kind of wonder, as if himself stupefied. 'That which, I suppose, induced him the more to encourage his superstition, was the continued series of contrary tempestuous winds, which had oppressed us ever since we had got into this sea...[he]

shot *the Albitross*, not doubting (perhaps) that we should have a fair wind after it.'

There's a fearful symmetry in the fact that albatrosses are now among the most endangered of all seabirds. They may spend a decade in the air never touching land, sleeping on the wing; like whales, they are able to nap with one hemisphere of their brain still active, in what is known, fittingly, as slow-wave sleep. And like the other procellarids, they also have an intense sense of smell. The ocean may seem a featureless expanse to us, but to an albatross it is a vast, multilayered web of odours. They can smell phytoplankton from miles away; and since the plankton's presence indicates invisible upwellings and seamounts, it not only leads them to sources of food, but allows them to navigate their way, orientating themselves to an olfactory seascape.

Yet even as they ride the updraughts on their exquisitely narrow wings, flying so free, they are susceptible to the activities of the human-altered world. Accidental death now haunts these birds, snagged by longlines or caught in nets. From the artificial islands of trash gathered by the Pacific gyre, albatross parents forage for bits of plastic that look like tasty morsels of squid and feed them to their chicks. Their bellies filled with drink loops, used cigarette lighters and tampon applicators, the chicks starve to death, leaving plastic-stuffed skeletons as modern *memento mori*.

As the albatross glides out of sight, we swerve into a feeding frenzy of short-tailed shearwaters, 'mutton birds' to the locals who still prize their oily flesh. Their teeming

shapes turn the sky into a living, squawking cloud, plucking at the sea with their wings. In his *History of Tasmania*, published in 1850, the Reverend John West reports on such immense flocks, as seen by Matthew Flinders, who first charted these waters in 1802. 'Captain Flinders…says that when near the north-west extremity of Van Diemen's Land he saw a stream…from fifty to eighty yards in depth, and of three hundred yards or more in breadth…during a full *hour-and-a-half* this stream of petrels continued to pass without interruption, at a rate little inferior to the swiftness of the pigeon. On the lowest computation he thought the number could not have been less than a hundred millions.'

Out of the bird-cloud comes a trumpeting blow, followed by a subtler whoosh, an intensely familiar sound to me: a humpback whale, travelling with its calf. They're stragglers in the spring migration, dawdling on their way to the feeding grounds – despite the mother's hunger, as evident from the vertebrae visible beneath her reduced blubber, having given all her energy to her calf. Her kind, like the southern rights, have been swimming this route for centuries. The Aboriginal people, to whom the sea was only an extension of their Dreaming, and who, like the whales, had no home, only a deep sense of their connection to their environment and a forty-thousand-year-old culture, would 'sing up' the whales, prompting them to breach in joy for the calves they were about to bear, or those they brought back with them. They also believed that when whales stranded themselves, they did so to feed human tribes in 'an act of courtesy, an act of promise'.

In ten years of whalewatching off Cape Cod, I've seen dozens of humpback mothers and calves. But there is something different about this pair. Perhaps it's the way they feed, lunging on their sides rather than gathering up their bait-fish in a net of bubbles. Or perhaps it's the female's flukes, much whiter than those of her northern counterparts. The image of the pale mother against the

black rocks is as stark and elusive as the solemn portraits of native people I'd seen in Hobart's museum; as if neither were really made for me to see. A promise, and a courtesy.

Just north of Adventure Bay, on the narrow neck that links North and South Bruny, is a sandy stretch where blue penguins waddle ashore at night to feed their chicks, running the gauntlet of the open beach to reach their young nesting in burrows in the dunes. High over this slender strand stands a stone cairn with the bronze relief of a woman's face; her name, Truganini; her dates, 1812–1876; and nothing else.

Truganini was the daughter of Mangana, chief of the people of Lunawanna-alonnah, as Bruny Island was known to her tribe. Their ancestors had lived there for thirty thousand years. But at the age of seventeen, Truganini watched her mother being stabbed to death by men from a whaling ship. Shortly after, sealers abducted her two sisters and took them to Kangaroo Island as slaves. Her brother was killed and her stepmother taken by escaped convicts. Then she and her betrothed were kidnapped by lumberjacks and taken to the mainland. During the crossing her husband-to-be was thrown overboard; as he tried to climb back on board, the men cut off his hands, leaving their victim to drown. Truganini was then repeatedly raped. Her shocked father, Mangana, died soon afterwards.

Truganini was witness to brutalities which the Reverend John West could hardly bear to iterate. 'If it were possible in a work like this to record but a tithe of the murders committed on these poor harmless creatures, it would make the reader's blood run cold at the bare recital,' he wrote, attaching a single footnote as if to spare the sensitive reader:

> *One case may suffice. A *respectable* young gentleman, who was out kangaroo hunting, in jumping over a dead tree, observed a black native crouched by the stones, as if to hide himself. The huntsman observing the white of the eye of the native, was induced to examine the prostrate being, and finding it only to be a native, he placed the muzzle of his piece to his breast and shot him dead on the spot. Hundreds of similar cases might be adduced.

One white man, however, appeared to have the native people's best interests at heart. On 30 March 1829, in the middle of the Black War between the settlers and the settled, George Robinson arrived on Bruny Island to create a new refuge, a new colony.

George Augustus Robinson, born in Lincolnshire in 1791, had emigrated to Australia in 1822. He set up a successful building company in Van Diemen's Land, but as a man of faith he was also drawn to good works, and with them, perhaps, social ascendency; even his friends found him pompous and vain, 'more patronizing than courteous and somewhat offensively polite than civil'. By that point the settlers' cruel treatment of the Aboriginal tribes – usurping their hunting grounds, kidnapping children and killing adults – had reached the point at which a final extermination was suggested, or at least the removal of these unwanted people from the lands which the settlers required. As a Christian, however, Robinson believed conciliation was possible. Appointed by George Arthur, he established a model village in order to redeem the Aboriginal people who, he had to admit, 'rank very low in the savage creation', yet were possessed of 'many amiable points which glitter like sunbeams through the shroud of darkness by which they are enveloped...'

In pursuit of his task, Robinson undertook a series of expeditions deep into western Tasmania, a place which even now remains a wilderness. With him he took a following of indigenous people, including Truganini, with whom, rumour suggested, he had formed a sexual relationship. Robinson attempted to persuade the people of these remote areas to join the proposed sanctuaries on Swan Island, Gun Carriage Island and Flinders Island, off the north coast of Tasmania. It was during this so-called Friendly Mission that Truganini saved Robinson's life when he was under attack on the Arthur River; the roles of the vaunted protector of the Aboriginal people and his childlike charges had become reversed. The story was widely reported in Hobart, thrusting Truganini's image into the public eye as the acceptable face of the savage.

Among her portraitists was Thomas Bock, himself a transportee (his crime had been to administer drugs to a young woman back in England). Now a well-known artist in Tasmania, he was commissioned by the new governor's wife, Lady Jane Franklin, to paint its native people. His work is poignant, since it shows how these people looked before their appearance was influenced by the coming

246

of the colonists. In his watercolour of 1837, Truganini appears as a young woman in her twenties, with shaven head and carefully arranged tribal dress; an Antipodean Eve, her breasts bound by a piece of twine. But in his oil painting, Benjamin Duterreau – born in London of French parents, he had emigrated to Van Diemen's Land in 1832 – created an evanescently beautiful, almost fairy-like portrait of Truganini, painted in recessive blue, green and brown against a crepuscular sky, as if she were already disappearing. It is an image only pretending to be real. Swathed in a kangaroo skin in lieu of ermine, with a shell necklace instead of pearls, she becomes a mythic, regal figure, like an Elizabethan princess in an alien land.

Was the artist painting his lover? She could be fourteen, or in her forties, this indigenous Mona Lisa. In his sketchbook, Duterreau added a note that summed up her rescue of Robinson in a similarly poetic, if not telegraphic manner:

Truggernana/A native of the southern part of V.D.land & Wife to Woureddy/was attach'd to the mission in 1829/Truggernana has render'd very essential service to the/expedition on many occasions & in a most remarkable manner/Saved Mr. Robinson's life by swimming & propelling/at the same time a small spear of wood to which Mr./ Robinson was clinging while endeavouring to cross/ the river Arthur to get away from some natives/who had form'd a plan to kill him. but not being/able to swim he owes his life to Truggernana.

The artist attempted to create a realistic portrait, but succeeded only in painting a dream. In his soft-focus and almost symbolist images, Duterreau depicted Tasmania's dwindling inhabitants in lush, highly composed and posed pictures. It seems fitting that his best-known work is incomplete, a study for a greater canvas. *The Conciliation* is a conversation-piece in the manner of Zoffany; Duterreau called it a 'national picture'. It shows George Robinson in his white duck trousers and navy cutaway coat and floppy cap, set against the elegant nakedness of the native people with whom he is portrayed in an unequal alliance, wagging his finger as if to lecture or admonish. He seems inappropriately, flamboyantly overdressed, while their loincloths are a modest invention by the painter. In reality, their naked bodies were protected from the elements by ochre and animal fat, and their heads decorated with dried mud. They are accompanied by the dogs brought by white men, now used by the Aboriginal people to kill kangaroos; one hound sniffs provocatively towards a grey wallaby.

As I look at it, in the parquet quiet of Hobart's museum, the canvas seems to transcend its history; what it meant at the time, what it means now. The entire arrangement has a mortal definition; it is divided by lines. The whiteness of its central figure asserts itself over the darkness of the others. Thin shafts of spears which could kill bisect the composition, giving it a modern tension. At its centre, next to Robinson like a dark mirror, is Truganini, pointing to her protector and lover, and to the future of her people.

But it was already too late. The island asylums to which they were led by Robinson were little better than penal colonies. His vision was betrayed by reality. In 1803, when it was first settled, there were about ten thousand Aborigines in Tasmania; by 1835, when Robinson took charge of Flinders Island, fewer than one hundred and fifty remained.

In their unnatural confinement, disease took its toll. Even the clothes they were forced to wear, partly to inhibit any attempt at escape, caused them to catch cold when

they got soaked in the rain, and as with Native Americans, fabric was the conduit of infection. 'Among savages, the blanket has sometimes slain more than the sword,' wrote the Reverend West. Some would bleed themselves to assuage their pain, blood streaming down their faces; it was the only way they could react to the helplessness of their fate. Others simply gave up the will to live. 'They were within sight of Tasmania, and as they beheld its not distant but forbidden shore, they were often deeply melancholy,' wrote West; he claimed that more than half those held on Flinders Island had died from '*home sick-ness*, a disease which is common to some Europeans, particularly in the Swiss soldiers'. Many starved themselves; healthy spouses who were bereaved 'would immediately sicken, and rapidly pine away'.

(Nor were their white visitors immune to such subtle suffering. In 1826 another of my distant cousins, Isaac Scott Nind, had sailed to Van Diemen's Land as an assistant surgeon in the 39th Dorsetshire regiment. Sent to the remote settlement of King George Sound in Western Australia, he became fascinated by the local tribes and later wrote a report on their culture for the Royal Geographical Society. But after five years in the wilderness, isolated with fifty others, two dozen of them convicts, Isaac was slowly going mad. He told the sergeant he'd rather see the back of a man than his face, and one day grabbed the commandant's hand and, in tears, began to blurt out an account of his past sins – then repeated his confession in the soldiers' barracks. Meanwhile, another man climbed a nearby mountain every Sunday to pray to

God for relief. Soon after, Isaac was sent home to England, suffering from a nervous breakdown.)

The settlers had no vocabulary or context for what they faced; that which they did not understand, they ignored, or destroyed. The Aboriginal people had both, but in their despair, seemed to prefer to die. In a charity shop in Surrey I once found a 1950s edition of James Frazer's *The Golden Bough*. Slipped into the pages on sympathetic magic, among accounts of Aboriginal people mimicking cockatoos and dugongs, I discovered a yellowing newspaper clipping which, to judge from its typeface, appeared to have come from a contemporary edition of *The Times*. The dateline was Darwin, April 16. 'A 19-year-old Aborigine who was said to have been put under a spell known as "singing him to death" by tribal women, was taken from an iron lung here to-day for 25 minutes and asked for food and water for the first time since he was admitted to hospital on Tuesday,' it reported. 'He cannot breathe voluntarily, and he cannot eat or drink. Doctors can find nothing organically wrong with him.'

Truganini's peers may have elected to die as passive victims, but she did not. She had tried to be part of the white world, or at least to work with it. In the aftermath of the Black War, during which the government had offered a bounty of £5 for each Aboriginal adult captured alive (and £2 for each child), Truganini moved to the mainland and joined a band of rebels, former whalers living on the outskirts of Melbourne, then a virtual war zone between the settlers and the settled. Resistance had become active.

In one notorious raid, two white whalers were murdered and other settlers shot. In the ensuing pursuit, Truganini was shot in the head. She survived, but was tried in court and narrowly avoided being the first Aboriginal woman to be hanged in Australia. Sent back to Flinders Island, she and the last of her people were then removed to Oyster Cove, south of Hobart. New images showed them bowed down by oceans of cheap fabric, engulfed in their captors' costume.

Shortly before her death from 'paralysis' in Hobart in 1876, Truganini confessed to a Church of England priest that she feared the dishonour of her body. She had good reason to do so, given the fate of William Lanne, her second husband. A well-known whaler, 'King Billy' had died in 1869, and his corpse had become the subject of an unseemly dispute between scientists of the Royal College of Surgeons, run by John Hunter's heirs, and those of the Royal Society of Tasmania, busy building their own collection. It is a remarkable story, as told by the historian Helen MacDonald, who points out that such grave-robbing had its precedent in the Antipodes.

In 1856 Joseph Barnard Davis, an English surgeon and collector, instructed Alfred Bock, son of the painter Thomas, on how to salvage specimens of Aboriginal people to supply a Western taste for such ethnic curiosities. In order to obscure acts of pillage which were, if not illegal, then certainly unethical, Bock was to find not just one but two corpses: one black, one white, ready for burial. Since his customers generally only wanted the cranium, he was to peel the skin from the black skull and replace it with that of the white person. He would then dress the replacement so that the skin would assume its shape. This ghoulish deception was, perhaps, the ultimate insult for a dreaming people: to go to one's grave with white bones beneath black skin.

Eight years later another Englishman sent a request for more bones. William Flower, who had succeeded Richard Owen as conservator at the Hunterian, was eager to acquire a sperm whale for the collection. He wrote to William Lodewyk Crowther, a Tasmanian surgeon and owner of a fleet of five whaleships, and was duly sent the skeleton of a fifty-one-foot male sperm whale caught off the south coast of the island in 1864, along with the lower jaw of the largest sperm whale ever taken in Tasmanian waters, measuring sixteen feet and indicating an animal of more than seventy feet in length.

But Flower had other acquisitions in mind, too. In a postscript to his letter he added the suggestion, 'I suppose there is no further chance of obtaining a skeleton of…one of the aboriginal human inhabitants, or a pair, male and female?' There was a hint of the black market in this whispered request. Crowther replied that there were just five such 'specimens' left – that is, Truganini and her friends at Oyster Cove – and promised to do his best. His grotesque efforts echoed John Hunter's 'collection' of the Irish Giant in the previous century.

On the night of William Lanne's death, Crowther and his son stole into the morgue in the Colonial Hospital. Employing the techniques outlined by Joseph Davis, they took Lanne's skull and replaced it with one from a white body. Did Lanne now look like a white black man, or a black white man? It hardly mattered, since he was no longer Lanne at all. Having discovered what had gone on, George Stokell, resident surgeon at the hospital and member of the Royal Society of Tasmania, the rival of its London equivalent, was told to amputate Lanne's hands and feet and so prevent Crowther from returning to claim the rest of the skeleton.

King Billy – or what was left of him – was carried to his grave by fellow whalers, among them a Hawaiian, a South Australian Aborigine and an African-American. His coffin was covered with a black opossum-skin rug, 'and followed by above a hundred citizens', as *The Times* noted, adding that Lanne was 'the last man of a race which only half a century ago numbered 7,000 souls...' But that night, Lanne's corpse suffered a third violation. Barely hours after interment, his coffin was dug up and his remains harvested by the Tasmanian scientists. His bones were as dispersed as St Oswald's, only for reasons of science rather than of faith. When the bizarre conspiracy became public, cartoons appeared in the press depicting Crowther as a grave-robber, surrounded by coffins dangling from ropes and bat-winged devils out of a Victorian pantomime. It was claimed he kept Lanne's skull as a paperweight on his desk in Hobart; its current whereabouts are unknown.

Given these gruesome events, Truganini's fears were entirely understandable. Shortly before she died, she asked that her body be cremated and her ashes scattered in the D'Entrecasteaux Channel that separates Bruny Island from Tasmania; like Charles Byrne, the Irish Giant, she sought the oblivion of the sea. Instead, she was buried in Hobart's suburbs, following a strange lying-in-state, wrapped in a rough red blanket and placed in a pauper's shoe-blacked coffin. Her friend, the barrister John Woodcock Graves, said he'd never seen a corpse so 'placid and beautifully quiet'. He suggested that a plaster cast be made of her face, and asked why the government had made no provision for the burial of such a notable person. Accordingly, the public were invited to 'Queen Truganini's' funeral – but only after the Royal Society in London had removed samples of her skin and hair.

Two years later, the Royal Society of Tasmania, which claimed, erroneously, that it did not have a female Aborigine specimen, was allowed to exhume Truganini's remains on condition that the skeleton was not put on display. Her bones were stored in a box until 1904, when they were articulated and put on show in a glass case in Hobart's museum as 'The Last of Her Race'. In 1947, a belated decency overcame the curators and Truganini was removed from the public gaze, but it was not until 1976, one hundred years after her death, that Truganini's final wishes were honoured: her remains were cremated and the ashes scattered off Bruny Island. As quickly as the events of her life had played out, so slowly history repaired the insult to her memory. In 2002 it was discovered that the samples of her hair and skin which had been taken were still held by the Royal College of Surgeons in Oxford. They were returned to Tasmania. Truganini had finally transcended her story. Her skeleton is gone from its glass cabinet, and does not even exist in the photograph that I have of it, which I cannot reproduce here.

On 21 April 1805 a letter from William Paterson, Lieutenant Governor of Van Diemen's Land, was published in the *Sydney Gazette*. It described an animal 'of a truly singular and nouvel description', which had been killed by dogs at Port Dalrymple, and which, so the excited editor informed his readers, 'must be considered of a species perfectly distinct from any of the animal creation hitherto known, and certainly the only powerful and terrific member of the carniverous [sic] and voracious tribe yet discovered on any part of New Holland or its adjacent Islands'.

Paterson's interests were scientific as well as military. He was another protégé of Joseph Banks, to whom he had dedicated his *Narrative of Four Journeys into the Country of the Hottentots and Caffraia*, written after his sojourn in South Africa in 1780. His report from Van Diemen's Land, which had the air of an academic paper, sealed the fate of the Tasmanian tiger. Like the island's peoples, its seals and its whales, such exposure would prove to be the tiger's undoing – and in the same short span. It all happened with extraordinary speed.

For thousands of years the thylacine, as it was more properly known, had survived in Tasmania long after its mainland population had been forced out, a result of an ever drier habitat and competition from dingoes. On this temperate, predator-free island – a residue of what Australia once was – it was preserved in its insular splendour, protected by the surrounding sea. Until now.

'It is very evident this species is destructive,' Paterson announced; and he had the evidence to prove it. 'On dissection his stomach was filled with a quantity of kangaroo, weighing 5lbs. the weight of the whole animal 45 lbs. From its interior structure it must be a brute particularly quick of digestion.' Guilty from the inside out, the creature's every part was measured and itemised: from its eye, 'remarkably large and black, 1¼ inches', to its tail, '1 foot 8 inches'. Paterson was nothing if not precise. He counted nineteen bristles on either side of the animal's face, and found its body to be covered with short smooth hair, 'of a greyish colour, the stripes black; the hair on the neck rather longer...the hair on the ears of a light brown colour, on the inside rather long...' Despite these anchoring details, one might have forgiven the reader of the *Sydney Gazette*, secure in his convict-built house in Port Jackson or Hobart, if he had assumed that this was a fabulous portmanteau of a beast, barely more credible than any of the other extraordinary fauna that leapt or crept or flew or swam around these islands. 'The form of the animal is that of a hyæna, at the same time strongly reminding the observer of the appearance of a low wolf dog. The lips do not appear to conceal the tusks.'

Such sins! The animal was betrayed by its features as much as any Victorian criminal by his photograph. The physiognomy was exact: unconcealing lips and low wolfishness, indicating cunning. Such strangeness added up to a sentence of death, and the thylacine, suspected of manifold offences, could hope for little mercy.

In 1808, George Prideaux Harris, Deputy Surveyor-General of Van Diemen's Land, sent Joseph Banks a sketch of another animal, which had been caught in a trap baited with kangaroo meat. He noted the tiger's 'near resemblance to the wolf or hyæna', and that its eyes were 'large and full, black, with a nictitant membrane, which gives the animal a savage and malicious appearance'. As newly discovered as it was, the thylacine was being coopted for the roles it had been assigned, according to the level of threat or scientific interest it evoked. Illustrating what he called the 'Zebra or Dog-faced Dasyurus' in an 1827 edition of his encyclopedic *The Animal Kingdom*, Georges Cuvier concluded, 'Its compressed tail seems to indicate that it is a swimmer, and it is known to be an inhabitant of the rocks on the seashore of Van Diemen's Land, and to feed on flesh...fish and insects.' In Cuvier's text, the thylacine resembles one of the early evolutionary mammals that took to the water to become entirely marine, the ancient ancestors of the whales, on whose descendants it now fed: 'They also seek, with avidity, the half-corrupted bodies of seals and cetaceous animals on the sea-shore.' Joseph Milligan, writing in 1853, agreed with this Darwinian demeanour: 'The Aborigines report that this animal is a most powerful swimmer; that in swimming he carries his tail extended, moving it as the dog often does.'

Gradually, more certain details began to accrue, bringing into focus a composite creation. Classified as *Thylacinus cynocephalus*, dog-headed pouched animal, the thylacine was one of only two marsupials – the other being the water opossum – in which both sexes had pouches, the male's covering its genitals to protect them as it ran in the bush (a refinement which is surely the envy of other males). It could raise itself on its back legs like a kangaroo, and ran at speed when hunting. Its elliptical pupils were perfected for night viewing, and its one-hundred-and-twenty-degree gape geared for scavenging. However, it would later be proved that the animal's jaw had little strength, and that its yawning gesture – often accompanied by the straightening of its tail and a peculiar strong scent – was not a sign of boredom, but a warning that it felt threatened and could be about to attack.

This fearsome reputation convinced settlers that the Tasmanian tiger was a danger to their stock. As early as 1850, the indefatigable John West was protesting its innocence, allowing that the thylacine did kill sheep, but only one at a time, unlike a wild dog or dingo, 'which both commit havoc in a single night'. Yet the Reverend had to accept that his defence was futile. Rewards offered by sheep-owners meant it was probable 'that in a very few years this animal, so highly interesting to the zoologist, will become extinct; it is now extremely rare, even in the wildest and least frequented parts of the island'. A male and a female were sent to the Zoological Society of London during the present year [1850], and were the first that ever reached Europe alive.' To many Tasmanians, it might as well have been an invention. Given its rarity, it is unlikely that my transported cousin James ever saw one, at least not in the wild.

While their peers were caged in foreign prisons, wild thylacines had bounties placed on their heads and hides. Five shillings was offered for a male, seven shillings for a female, with or without pups, and from 1878 to 1909 more than four thousand thylacines were culled. Some became

waistcoats and rugs, allowing men to both wear and walk
on their trophies. The animal's steep decline was only
accelerated by the reduction of its habitat, a distemper-
like disease, and predation by domestic dogs. By 1910 the
population was scarce; the last confirmed thylacine to be
killed in the wild was shot in 1930 by a farmer, Wilfred
Batty. A photograph of the carcase propped up against a
fence displays the power of this creature even in death.
The farmer's sheepdog backs away in fear, flecks of
spittle around its mouth. The last capture of a thylacine
took place in the Florentine Valley in 1933. Thereafter

all is supposition, although in 1946 Dr David Fleay came close to trapping a thylacine on his expedition into the Tasmanian interior. And that was the final encounter – as far as science is concerned.

In Hobart's museum, next to a gallery lined with Duterreau's paintings of Tasmanian Aborigines, a video plays on a constant loop. It shows a female thylacine that was bought by London Zoo in January 1926, and which would die on 9 August 1931, shortly after the film was made. It also displays footage from Hobart's zoo, where the Florentine thylacine had been brought. This hapless beast was a victim of human circumstance twice over, since at that time Tasmania was subject, like much of the Western world, to economic depression. In the mid-1930s, Beaumaris Zoo ('beautiful marsh') had become run-down, and its inhabitants – among them polar bears and elephants – neglected in their concrete compounds.

Like the stranded sperm whales of the Mediterranean, every factor seemed to conspire to precipitate this last captive's demise, as Robert Paddle, a doctor of psychology at the Australian Catholic University, writes. The deciduous tree which had covered the thylacine's cage had shed its leaves. 'Without access to her den, the thylacine was unshaded from the extreme, unseasonal heat by day, and shelterless from the extreme cold by night. Thus, unprotected and exposed, the last known thylacine whimpered away during the night of 7 September 1936.' It is a mournful scene, worthy of a Victorian oil painting: the bare tree, the freezing night, the crossed paws as the forgotten tiger lays down her head. Like the Aboriginal people, she too had pined away, homesick in her own homeland: collected, and abandoned. A year later, the zoo closed, and the site became a fuel store for the Australian navy.

But in the technology of the twenty-first century, the thylacine lives on. The Hobart and London films are remarkable for the tantalising glimpses they allow us of an animal so recently made extinct. Their subjects resemble flickering ghosts, pacing up and down, caught in a loop, yet their physicality is certainly not spectral. In bright sunlight, the Hobart specimen is seen in precise detail. It is a living chimera. The mountain-lion head. The narrow, almost cetacean jaw, yawning as if to display distress at its observation. The lemur-like stripes on its hindquarters and the kangarooish tail, thick at the base and tapering to a whip-like point and which, I guess, must have felt as heavy as the wallaby's tail I once weighed in my hands while its owner foraged on the ground.

In other photographs of a thylacine family group at Beaumaris in 1909, taken only a century after Paterson first described the animals, they appear intimidated by the photographer's lens. The vividness of these images renders their subjects almost domestic; I imagine finding one curled at my feet – not such a far-fetched idea, since captive thylacines were often given collars and walked on leads like dogs.

From the 1880s to the 1920s, a total of nearly two dozen thylacines were sent to London Zoo, a hub for such exotic specimens, an imperial animal clearing house. Many went on to other compounds, in New York and Berlin. In captivity, these unassertively, subtly strange creatures did not speak up for themselves. 'It is unfortunate that only rarely did anyone take time to observe them,' writes one modern commentator; 'their tranquil nature did not arouse much interest in zoo visitors, or zoo directors, either.' Their calls were said to resemble the slow opening of a door, but their cells would stay shut.

It was only in 1986, fifty years after the death of the last known specimen, that the thylacine was declared lost to the world. Even now CITES qualifies the tiger's status as only 'possibly extinct'. Given such a recent extirpation in an island as untamed as Tasmania, it was inevitable that stories would persist of its survival.

In 1957, a photograph was taken from a helicopter flying over Birthday, on the west of the island, of a 'striped beast on the deserted beach'. 'It was probably a thylacine,' the Belgian naturalist and cryptozoologist Bernard Heuvlemans claimed, matter-of-factly. 'An expedition was at once mounted in order to capture a specimen, which would be released again after it had been studied.'

Heuvlemans' confidence was misplaced. Despite other reports of tracks and attacks, and the best efforts of a Disney film crew and an expedition led by Sir Edmund Hillary in 1960, no specimen was found. However, in 1961 a pair of fishermen apparently came close to capturing one by accident when it was snared in a trap. The two men, Bill Morrison and Laurie Thompson, risked ridicule to talk to the Hobart *Mercury*. 'The tail was rigid,' said Morrison. 'The animal's coat was dark, and I could discern only one stripe behind the shoulders and extending around the chest.' The beast was understandably maddened, reported the newspaper, although the sound it made was 'rather peculiar, and different from the barking of a dog'. As the two men attempted to release it, it escaped – although not at any great speed. 'It seemed to be a slow mover,' said Morrison.

The thylacine had become a shadow of the past projected onto the present. Did it still exist, or not? In 1966, a six-hundred-thousand-hectare game reserve was set up in south-western Tasmania, partly to protect any animals that might remain in the area. It was both a futile and an optimistic gesture, as if a space had been made, ready to be restocked with all the flora and fauna that had disappeared from the island.

Since extinct animals no longer exist, we must take their once-existence on the word of others, especially if they seemed improbable in the first place. Even living animals defy our comprehension. The narwhal, for instance, with its icicle-like and onerous tusk, seems too strange to have survived into the twenty-first century, and yet, having never seen one, I take its existence on trust. Equally, the tiger-striped stuffed dogs with kangaroo legs which I saw in Hobart's museum might well be clever fakes, just as the first duck-billed platypus brought back to Europe was declared a preposterous and obvious forgery.

Set back from the city's Parks Road, Oxford's University Museum was built, from 1855 to 1860, on what was then an open plain north of the city. Despite its splendid stone and slate façade, it is an only half-finished building. It was designed by Benjamin Woodward, a civil engineer from Cork with a fondness for medievalism, but its presiding geniuses were John Ruskin and his friend Sir Henry Acland, who determined that here, on the university plain, art and science would meet in a glorious hall of knowledge. 'I hope to be able to get Millais and Rossetti to design flower and beast borders,' Ruskin told Acland, '– crocodiles and various vermin – such as you are particularly fond of – Mrs Buckland's "dabby things" – and we will carve them and inlay them with Cornish serpentine all about the windows. I will pay for a good deal myself, and I doubt not to have funds. *Such* capitals as we will have!'

The new museum would bestride the Oxford lawns, its iron and glass roof wrought in the image of plants and animals, and its cast-iron pillars topped with stone capitals and bases sculpted with fork-leafed and serrated ferns and fox-like animals peeking in between. Others sprouted lilies with protruding stamens, waiting for some masonry bees to pollinate them.

At the entrance to the museum, and almost too big to be seen, hanging by a crude iron hinge and standing at more than double the height of a man, is the lower jaw of a sperm whale. Splayed like an enormous wishbone, its broad, veined bone narrows and curves to meet in a gothic arch, tall enough to create its own, alternative entrance to the museum; little wonder Melville called such a jaw a 'terrific portcullis'. From under its shadow, I am conducted by the smartly dressed Malgosia Nowak-Kemp, who invites me, in her faint Slavonic accent, to follow her along a corridor, over a thick red velvet rope, and through a heavy wooden door. The dark, cave-like interior beyond is filled with modern, white-painted metal shelves which, as Malgosia presses a discreet button, slide apart to reveal row upon row of glass jars.

Here is the usual array of mammalian organs and dead-eyed fish, the everyday horrors of the half-known world. A monkey's head sits on top of a dissected spine, like something from an etching by Odilon Redon. But in a corner on the floor by the shelves stands a tall transparent cylinder, about waist-height, with an unsecured glass disc for a lid. The liquid it contains is the colour of stewed tea. And in it, held upright by a white rag tied around its neck, giving it the appearance of a convict who might escape at any moment, is a thylacine.

Confined in its glassy prison, the animal is difficult to distinguish. It might be an oversized rat. I peer in at the sides and over the top, trying to imagine its living magnificence. From one of the shelves, Malgosia fetches a smaller jar. In it is thrust the skinned head of another thylacine, its skull removed, the deboned face swirling in its alcohol like a pickled glove. The fur is that of a dog or a fox, Labrador-pale and in good condition, as though someone had preserved a much-loved family pet.

Back in her office on the far side of the building, Malgosia opens the wooden doors of a cupboard to another pair of specimens: busts of Truganini and Woorrady. Briefly, dramatically revealed, they stand blank-eyed on their pedestals in the shadows, simulacra of their countrymen and women whose bones lie in other cupboards and drawers of this museum, awaiting repatriation.

In an anteroom, perched on top of a set of dusty metal-framed shelves as though it leapt up there one day and was afraid to come back down, is a stuffed thylacine, donated to the museum by the Royal Society of Tasmania in 1910. When Malgosia steps out of the room, I illicitly climb up the ladder-like shelves to unveil the marsupial from its inelegant plastic wrapping. It's fixed to a wooden base in an upright position, like the toy dog on wheels my sister pushed about as a child. Its fur is threadbare, and its lips have been sealed by the taxidermist's art. Having photographed it, half hanging onto the shelves, I reach out and stroke its leg, much as I once surreptitiously stroked the behind of a dozy koala in Sydney's zoo. Had I attempted such an intimacy with a living thylacine, the result might have been quite different. When Dr Fleay attempted to photograph the last tiger in Hobart's zoo, he was rewarded with a snap of its jaws; another visitor was bitten on the buttocks. Yet this stiff semblance of an animal is charged with a mystique; perhaps it will rub off on me, just as when I once shook the hand of a man who'd shaken the hand of a man who'd shaken the hand of Oscar Wilde.

In the shoebox-sized study room, I look through the careful notes and drawings made by Professor Tucker, who dissected the thylacine from London Zoo – or, at least, its head – in 1942. The world might have been at war, but these learned men were exchanging letters over soon-to-be extinct marsupials.

Oxford Museum of Natural History

5 Aug 1942

Dear Tucker,

Huxley has passed to me your letter of the 3rd inst,
re. Thylacine material etc. I regret we cannot supply
the trunk of the Thylacine, the head of which you
are now dissecting, because the remainder of the
animal was sent on Oct 8th 1931 by order of the
Society, to Professor Rowan of the Natural History
Museum, Edmonton, Canada, in exchange for a
collection of live stock Rowan sent to the London
Zoo! We have had no other Thylacine and it is
doubtful if we shall ever get another...

Yours very truly

A.E. Hamerton.

The Huxley to whom Hamerton referred was Julian,
brother of Aldous and grandson of Thomas Henry Huxley,
'Mr Darwin's Bulldog'. He was Secretary to the Zoological
Society of London, and that year was also corresponding
with T. H. White in Ireland on the subject of whether animals had a 'mind'.

Spilling over the desk, page after foolscap page details
every aspect of the animal's anatomy with exhaustive
descriptions written in Professor William Tucker's neat
hand. His is an update of William Paterson's report in 1805;
and as Paterson's had been the first detailed description of
the thylacine, so Tucker's might be the final one. As far as
the professor knew, he was the last person to have direct

physical contact with the extinct animal. It was as though he had been able to perform a necropsy on a velociraptor.

Putting the professor's minutiae to one side, I turn to another document in the file, one which purports to summarise thylacine sightings in north-eastern Tasmania in the last two decades of the twentieth century. Assembled by a husband-and-wife team, it is a painstakingly compiled list of encounters experienced by people used to the Tasmanian bush, one of the most protected wildernesses left on earth. These witnesses are well qualified and reliable, the report is eager to stress, and include practised bushmen and retired university lecturers. They were going about their ordinary lives when something extraordinary interrupted them.

One couple, driving back one night after having been to the movies in Launceston, saw a pair of strange shapes amble across the road. At first they mistook them for dogs. But as the animals were caught in the headlights, they saw erect ears on large heads, unlike any canine. The pair moved slowly, even nonchalantly, said the witnesses: 'It was almost as if they were disdainful of the car'; as if it, not they, were the interlopers. Amazed by the sighting, the couple reported it to the Parks and Wildlife Department, where an apparently uninterested official heard them out, then said, 'Yes, it looks like you saw what you saw. Now, will you do us a favour and shut up about it? Don't tell anyone.'

Such encounters stress the odd motion of the animals. One experienced bushman in his fifties was logging in the forest with three other men when an animal that corresponded to a thylacine – 'Couldn't have been anything else' – ambled out of a tree, 'as slow as you please… He wasn't in a hurry. But, then, they aren't very fast, anyhow.' All four men saw it for long enough to observe its strange gait and the rigidness of its hindquarters, the way that it couldn't turn around like a dog because of the stiffness of its back, but instead had to move in a circle. And in 1980, a woman in her own garden found herself face to face with a creature which she too identified as a thylacine. It was standing on her chicken coop. 'It stared at me and I stared at it. It was really quite beautiful. Sort of golden. It had a big head and stripes across the base of its rump.' In this brief moment, both were transfixed. 'We just sort of stared at each other.' She called quietly to her husband inside, at which point the animal disappeared. When she went back inside, she was grey and shaking.

Many witnesses remarked on the animal's serenity and stillness; 'No wonder they got killed.' Others claimed to have smelled its pervasive aroma, described a century ago as that of an unknown herb. Some came close enough to look into its large yellow eyes. Nor are these sightings confined to Tasmania: others have been reported from the Australian mainland, where video cameras captured dog-like animals ambling through the bush. Yet nothing definitive exists, only blurry smears that may or may not be anything other than a mongrel and yet which, like the 'true' films of the tiger, take on a strangeness all of their own.

One sequence, shot in 1973 through a windscreen (the wipers occasionally get in the way), is perhaps the most persuasive, since it is a mixture of the banal and the potentially astounding. An animal runs out of the trees and across a road. It might be a wild dog, but it has a long, stiff and pointed tail. It is powered by thickly-muscled back legs, which look more kangaroo-like than canine, and in the sunlight, magical stripes appear across its back. Rerun in slow motion, the images seem at once both eerie and ordinary, something caught between worlds.

In 1982 Hans Naarding, an experienced field ranger with the Tasmanian Parks and Wildlife Department, was in north-west Tasmania, conducting a survey of the Latham's snipe, an endangered migratory bird. He'd been sleeping in his vehicle when he awoke to heavy rain.

It was two o'clock in the morning. Out of habit, Naarding scanned the bush with his spotlight. 'As I swept the beam around, it came to rest on a large thylacine, standing side-on some six to seven metres distant.' The ranger's camera was out of reach – sceptics might say it always is – but anyway, he didn't want to disturb the animal. His decision allowed him to make detailed, and convincing, observations. 'It was an adult male in excellent condition with twelve black stripes on a sandy coat. Eye reflection was pale yellow. It moved only once, opening its jaw and showing its teeth.' Having watched it for several minutes, Naarding took his chance and reached for his camera. As he did so, the animal moved off into the undergrowth, leaving a strong scent in its wake. Because of his professional position, Naarding's sighting was taken seriously. It was also kept quiet while an intensive two-year search was made over two hundred and fifty square kilometres. Nothing was found.

Even as I write, Dr Stephen Sleightholme, who has made a lifetime study of the thylacine, shares with me a message he has just received from a witness who, a few weeks before, had apparently watched a Tasmanian tiger at eight o'clock in the morning, in broad daylight. As Dr Sleightholme notes, the thylacine was a shy animal that preferred the twilight; even when it was relatively common, it was rarely seen in any great numbers – it was no pack-running, predatory wolf. But perhaps the most intriguing evidence for its putative survival is supplied by dry statistics. In the early 1990s, Professor Henry Nix of the Australian National University developed a computer-generated map to correlate recent sightings, using a programme, BIOCLIM, created to predict where specific flora, fauna or ecosystems should occur.

Professor Nix used this map to compare historical records of thylacines hunted or trapped in Tasmania during the late nineteenth and early twentieth centuries with the frequency and location of sightings from 1936 onwards. The two sets of data coincided almost exactly, leading the professor to conclude that these witnesses might indeed be seeing thylacines, and he proposed that an official search should be made before going ahead with the considerable expense of attempts to clone the species from its DNA.

As to what all this adds up to, I do not know. There are always hoaxes and misidentifications and rumours of conspiracies and vested interests. Indeed, if a living thylacine were to wander out of the bush and be wrestled to the ground – as one Tasmanian professor of zoology fantasised – it might mean an end to the exploitation of the island's virgin forests, an industry that angers many Tasmanians as they see ancient trees being cut down and shipped out to make toilet paper.

What I do know is that in one institution I visit, a curator lets slip a quickly retracted remark, telling me it is not their secret to reveal. It is clear from what this person says, or does not say, that this strange half-life limbo of an animal which may or may not exist may soon be resolved, in its favour. That history is about to be reversed. That the thylacine is no longer extinct.

If it ever was.

7
The wandering sea

Far from land, far from the trade routes,
In an unbroken dream-time
Of penguin and whale,
The seas sigh to themselves
Reliving the days before the days of sail.

DEREK MAHON, 'The Banished Gods'

T ake out your atlas and look at it.

You can't. Just as no two-dimensional map of the world represents the true proportions of its continental masses, so no chart presents the reality of its greatest ocean. If you were to rise up, like an impossibly elevated albatross, from the centre of the Pacific, the Earth would appear almost entirely blue. No wonder Arthur C. Clarke thought a better name for our planet would be the Sea.

The facts defy that paltry layer of land which we call home. The Pacific contains one hundred and seventy million cubic miles of water, and covers sixty-three million square miles, a third of the planet's surface. It descends to the greatest depths, the Mariana Trench, nearly seven miles down, a place visited only twice by human beings, a lightless sea bed less mapped than the moon. This ocean also holds the oldest water: such are its slow-moving undercurrents that the oxygen in its middle layer has been there since that water was last in contact with the air at the surface up to a thousand years ago. It is as much an archive of the ancient atmosphere as the air bubbles trapped in ice cores taken from the Antarctic. Given these superlatives and immensities, it is not hard to believe that of the million species in the sea, three-quarters are yet to be described, and one third remain unknown to science. Even its sepulchral abysses support life forms alien to our imaginations: colourless creatures, living far from the sun and the photic layer, feeding off another energy entirely from volcanic vents. Life itself may have begun in such places.

Yet the Pacific is by no means a landless, uninhabited expanse. It is studded with twenty-five thousand islands, large and small, each with its own stories, of people, and animals. Their narratives crisscross the ocean in an embroidered web, drawn together in invisible lines of connection from shore to island to sea, transversed in ancient feats of navigation and migration that put our modern, computer-assisted efforts to shame. Here the remotest journeys ended; and here many began, too.

Out of the silence and darkness of the night, I'm plunged into a hubbub of people and cars and cargo, all jostling to join the ferry that rises from the quayside. An hour later, and the ship is pushing out from harbour. The sun seeps back into the sky, and the ocean opens up to meet us. I settle on the top deck to drink tea, lazily raising my binoculars – only to see a huge grey shape in the mid-distance. It takes me a moment to realise that it is a whale.

A voice crackles over the Tannoy to alert the passengers to the sight. They lurch over the rail for a better look. The sperm whale's blow fizzes in the air. It raises its head, shiny with seawater, but it doesn't really look like a whale. The passengers soon lose interest, and drift back to their breakfasts.

I watch as the whale slips into the distance and, with a final flourish of its flukes, dives. Such a sight is rare in these waters nowadays. Yet a century ago, ships plying this route were often accompanied by another cetacean – one which acquired a near-mythical status.

From 1888 to 1912, a Risso's dolphin appeared regularly in the Cook Strait, the turbulent channel that separates New Zealand's two islands. Nicknamed Pelorus Jack after Pelorus Sound, it was seen by thousands of passengers, including Rudyard Kipling and Mark Twain. Some claimed it as a kind of guardian angel, guiding ships across the dangerous waters, but it is more likely that the animal was surfing the compression wave created by the vessels' bows, as many dolphins do. A third theory suggested a more emotional connection: that it had lost its mother and was seeking a surrogate – an idea encouraged by accounts of unweaned cetaceans attempting to suckle at the sides of the same whaleships that had made them orphans.

Other remarkable abilities were claimed for Jack. That he could choose between two ships as to which would make a better scratching-post to rid his body of parasites, and that he preferred to follow steamers because of the sound they made. He even attracted the attention of London's Linnean Society, whose president, Sidney Harmer, the director of the Natural History Museum, noted: 'In the light of this story, we may have to review our incredulity in regard to the classical narratives of the friendliness of dolphins towards mankind.' But Pelorus Jack would be both endangered by and rewarded for his dalliance with humans. After a drunken passenger on the ferry SS *Penguin* took a pot shot at it in 1904, the dolphin became the first marine mammal to become protected by law: a hundred-pound fine awaited anyone who interfered with it. It was reported that Jack declined to escort *Penguin* thereafter; five years later the ship was wrecked off the South Island, with the loss of more than seventy lives.

For the native people of New Zealand, Pelorus Jack evoked an older myth, one which reflected their ancient relationship with the islands they called Aotearoa, the land of the long white cloud, a place shaped as much out of memory as from rock. To them, whales and dolphins were *taniwha*, shape-shifting spirits, and Jack was one in a long line of such animals to assist the human race. In the founding myth of their nation, a young man, Paikea, is nearly drowned by his jealous brother before the whale Tahoa appears and carries him on its back to Aotearoa. And while the West still saw cetaceans as monsters at the edge of the world, as living islands or spouting sea dragons, here at the real end of the world – according to occidental projections – their true nature was better known.

As a maritime people, the Māori were familiar with whales and birds and their movements. Attuned to the changing colour of the water and the direction of the prevailing winds, they navigated using their bodies; men even used their swinging testicles to sense the sea's swell. The Polynesians' first migrations followed those of cetaceans – what their Anglo-Saxon seafaring comrades called *hwælweg*, 'the whale's roads'. Even their physical attributes seemed to reflect one another: the islanders' broad, muscular bodies – so valued in the rugby players that they export – provided the power to paddle their canoes, while generous body fat sustained them, like blubber, on those long voyages.

To ally oneself to a whale is not so strange; some might say it is perfectly reasonable. Throughout history humans have celebrated their animal affiliations. Earl Siward of Northumbria, an eleventh-century warrior who carried a raven banner and defeated Macbeth in battle, claimed descent from a polar bear; his father's ears were said to be distinctly ursine. In my *Children's Hereward*, a book I was presented with at primary school 'for pleasing progress', the flaxen-haired, handsome young Saxon hero fights a mighty white bear, 'said to be of magic birth, and...related to the great Earl Siward himself', which is kept caged in a courtyard along with other wild beasts. When the bear escapes, kills a dog and threatens a terrified maiden, Hereward leaps from his horse and, to his own astonishment, slays the animal. The medieval world entertained such cross-breeds: men with stags' heads or trees growing out of their mouths, women with fishes' tails; chimera caught between magic and science. Even evolution, in its fluidity, appeared to allow for these hybrids: witness the thylacine, or Darwin's speculation that the fish-eating bears of the north-west Pacific coast might become entirely aquatic, 'till a creature was produced as monstrous as a whale', although he later regretted his flight of fancy.

The scientist was only, if unconsciously, reflecting the beliefs of Northwestern Indians, for whom the world was torn between terror and beauty and who lived on the edge of the sea because they found the land more fearful. Their carved wooden figures, preserved in Vancouver's airy Museum of Anthropology – where totem poles reach up to the glass roof and threaten to burst out of it like redwood trees – might as well be anatomical displays in the Hunterian. Outsized otters and mash-ups of whales and wolves blur reality with unsettling amalgams of claws and coiled tails. In one fantastical carving, multiple dorsal fins poke out of one lupine body, as if there were six whales inside, trying to break out. Meanwhile, over them all hovers the trickster Raven, mating with an oyster and delighted to discover, nine months later, that it had spawned mewling men to be let loose on the world.

The vast Pacific, which still seems so remote to the modern-day Western world, invoked such magical animal–human affinities. Its aboriginal cultures even seemed similar, as they reached from one coast to another. On a boundless, restive sea belied by its name, anything could become anything else. As Jonathan Raban wrote in *Passage to Juneau*, long before the white men arrived at the north-west Pacific shore – in journeys that connected Cook and Vancouver to the Antipodes – the native people had known what to expect from the flotsam washed up on their shores: bits of ships studded with nails that indicated an alien technology, much as if a modern beachcomber had found parts of a flying saucer.

The Māori's arrival on Aotearoa only underlined the importance of its animals, especially whales, in a land that lacked any native mammals. Like Tasmania, these were ancient islands, with their own unique, pre-human populations. New Zealand was formed out of the super-continent of Pangea, from which it had broken away sixty-five million years ago. Its only quadrupeds were reptiles, its largest animals, birds; and it was all the more Edenic for its dearth of fearsome predators. In such a place cetaceans were an important source of protein. (Western visitors would assume that the islands' natives resorted to canni-balism out of that lack of flesh.) And while Europeans were still calling whales fish, the Māori had a long-established taxonomy for the species they knew intimately, that they both used and venerated. It was an alternative classifica-tion, created centuries before Carl Linnaeus had begun to itemise the world.

Tohorā was the general name for whales, but also signified southern right whales. *Hakurā* or *iheihe* were scamperdown or beaked whales, many species of which swam in these deep waters; *paikea* was the humpback, *pakake* the minke, *ūpokohue* the pilot, and *parāoa* the sperm whale. Whale tribes had honorific titles, too, some-how more evocative, in their unfamiliar consonants and vowels, of the whales' strange beauty than the ugly names Europe had bestowed on them: *Tūtarakauika, te Kauika Tangaroa, Wehengakāuki, Ruamano, Taniwha, Tū-te-raki-hau-noa.*

For the Māori there was no demarcation between the life of the land and that of the ocean; such distinctions made no sense. Trees and whales were as one. The god Te Hāpuku was ancestor of both whales and tree ferns, known as fish of the forest. As medieval bestiaries drew correspondences between animals on land and in the sea – the elephant and the whale, the wolf as a shark, the goose born of barnacles – so the Māori saw the sperm whale in the kauri tree, a podocarp that grows to a hundred feet or more and can live for thousands of years. They related that when the *tohorā*, or whale, asked the kauri to accompany him on his return to the ocean, the tree preferred to stay on the land. Instead, they shared skins. Hence the thinness of kauri's bark, as oily as the whale's blubber, both wrinkled in age and majesty.

Humans too were interchangeable with whales. *Te kāhui parāoa* meant a gathering of sperm whales, but also a group of chiefs. *He paenga pakake* or beached whales indicated fallen warriors on a battlefield, while men assumed the guise of whales in their warfare. The Ngāti Kurī tribe created a Trojan whale from dog skins in which were hidden one hundred warriors; when their besieged enemy came out to feast on its meat, they were killed and themselves eaten. Other warriors lay on the beach in black cloaks to lure those who thought they'd found a pod of *ūpokohue*. And the greatest of all chiefs, Te Rauparaha, sustained his army with blackfish that had been driven ashore and tethered by their tails using strong flax ropes, to be killed as required, like a living larder.

Like Australian Aborigines, the Māori did not actively hunt whales, but made good use of stranded animals. Unlike Westerners, they did not render the blubber into oil and discard the rest; the entire animal was a resource which could provide for the tribe. The meat was eaten immediately or dried for later use, and they drank the milk from nursing mothers. Whale oil supplied polish and scent. Teeth and bone became adornments, the most precious being the *rei puta*, a whale-tooth pendant. The sperm whales' hard, dense bones also made broad blades and clubs that bore the power of the animal that had provided them.

Hundreds if not thousands of whales still beach themselves on the shores of New Zealand every year, and are regarded as *tapu*, sacred signs. When a pod of pilot whales stranded on the South Island recently, a Māori elder arrived with his sleeping bag to spend the night with them in order that they should not die alone. Solemn blessings are given to dead or dying whales. In one famous incident in 1970, fifty-nine stranded sperm whales were declared to be *tangata*, or human, and were interred in a communal grave, five hundred feet long. Their deaths were, paradoxically, seen as a good omen for an imminent visit from the Queen (and an Antipodean reflection of the medieval right to 'royal fish'). The same incident also inspired Witi Ihimaera's novel *The Whale Rider* – although his book was born in New York.

In 1985 the writer was working as a diplomat in Manhattan when a humpback whale swam up the Hudson as far as 57th Street. It seemed to Ihimaera to have come up 'that dirty big black primordial river' to see him, as one emissary to another. 'I have never believed that the Māori world stops when you leave the country,' he told me, 'nor have I ever believed that the interconnectedness – that interface as you call it – stops simply because it's dysfunctional now on the human side. Do whales have ancient memories? Sure they do.' The New York whale represented all the whales that had been so important to his people; it was a symbol of the ineluctable past, the present and the future.

Orphan or guardian, Pelorus Jack was last seen in 1912, and probably died that year of old age, although some suspected that a visiting fleet of Norwegian whalers had harpooned him. Nowadays his legend is reduced to a company logo for the ferry line, a cute dolphin leaping over a wave in the shape of the national fern leaf. There is no need for cetacean guides now. As our ship reaches the narrow fjords of South Island, where a Nordic whaler would feel at home, it steers to signals bounced from unseen satellites. The tree-clad slopes plunge straight into the sea. It might almost be a pretty scene, with its fluttering yachts, were it not for the sense that the cliffs are closing in, as if to squeeze us in their grasp.

From the deck below me, I hear sheep bleating, caged in a truck.

At the end of Queen Charlotte Sound stands the town of Picton, itself a former whaling port. The ferry inches into dock, and I step out onto the street and walk over to the railway platform. The rackety train leaves on the dot of one, rattling through green valleys studded with vineyards. The last carriage is an open-sided observation car; I feel like a character from the Wild West as I lean out into the wind. We pass salt flats blushed pink by algae and sheep grazing in fertile pastures, negotiating the narrow corridor formed by the Southern Alps to the west, and the open Pacific to the east. Forced ever further south by their twin wildernesses, we arrive at Kaikoura. This former railway town was once busy processing sheep. Now it advertises its new fortune in a punning sign over the old ticket office: the Whale Way Station. But a wobbly horizon says no more boats today.

In a bar on the town's main street at what he calls beertime, I meet Bill Morris, who has driven all the way from Dunedin to talk whales. He's tall, in his twenties; his hair stands up in tufts from sleeping in his camper van. We climb into the front seats, with his clothes and guitars piled in the back, and drive down to the headland.

The landscape is so sharp and bright and brutal that it's as if I'd administered eyedrops. At low tide, the plateau of rock is covered with seaweed of every shape and colour, a vivid herbaceous border of bladder-wrack and coral-like fans. Bill and I walk out to the water's edge, through swathes of bull kelp laid flat like a felled forest. New Zealand fur seals loll, their flashing eyes daring us to get closer. Their name, *Arctocephalus forsteri*, commemorates Georg Forster, who with his father Johann Reinhold Forster took Banks's place as naturalists on Cook's second voyage in the 1770s. Georg Forster brought back a sketch of what he called sea bears, an augury of Darwin's confusion, as if they'd fused their paws into flippers and taken the plunge into the sea. By naming the beast Forster laid it open to its fate, as Paterson had done for the thylacine, as the *Discovery* scientists would do for the whales of the Southern Ocean. In just one season in 1824, eighty thousand fur seal pelts were taken from South Island. Their numbers remain a fraction of what they once were.

Back on dry land Bill shows me a lonely, pink-painted wooden hut with a tin roof. Whale ribs are scattered around its stoop; its foundations stand on vertebrae. Just as whales brought its first people here, this country's foundation, as far as the West was concerned, lay in whaling. This tiny cottage was once part of the station established here in 1842 by Robert Fyffe. He and other whalers took up to fourteen thousand southern rights each year; to them the stink of rotting carcases was 'the smell of money'. By the 1920s, ninety-nine per cent of the Southern Hemisphere's right whales had been killed. All that remains of the factory is a single fireplace, standing like a wayside shrine on the shore, as if everything else had been washed away.

Inside the cottage, the cabin-like rooms are still covered in Edwardian wallpaper. The low southern sun forces through the windows. It feels like the last house in the world. There's a sense of extinction and improbable life; of enormous animals and their equally vast absence. Here you might not be surprised to see a giant bird stalking through its cottage garden, and indeed, four centuries ago you might have seen just that. The largest moa egg ever found was discovered here, evidence of an earlier cull carried out on this shore.

The biggest of the moa species could reach twelve feet, twice the height of a man, but that did not prevent them from being hunted to extinction. By the beginning of the sixteenth century, while Tudor monarchs were squabbling in England, the moa had retreated to New Zealand's South Island, just as the thylacine made its last stand in Tasmania. Here the most complete remains have been found, including the fearsome claws of *Megalapteryx didinus*, discovered in an Otago cave in 1878 with its feathers intact, like the hairy remains of giant sloths found in South American caves. In this sublime, scaled-up landscape, it is hard to believe that such large creatures disappeared so recently; their shapes seem to linger, like the after-image of the sun.

In 1844, Kawane Paipai, a Māori elder, told Robert FitzRoy, who had been captain of the *Beagle* during Darwin's voyage and had subsequently become governor of New Zealand, of a moa hunt that had taken place on South Island fifty years before. In a scene that recalls one of Ray Harryhausen's films, Paipai remembered the bird being harried and surrounded before it was speared, using weapons that, like whaling harpoons which bent as they entered the blubber, were constructed to snap once they struck. A frightened moa would fight back, using its huge feet to strike at its attackers – although this tactic left the bird unbalanced and easily toppled from behind. Yet more cruelly, other birds were killed by being made to swallow hot rocks.

Despite such depredations, many were convinced of the moa's survival into the modern era. Whalers and sealers said they saw monstrous birds on the rocky shores. Bones were discovered with marks that, it was claimed, could only have been made by iron blades unavailable to the Māori – indicating that Europeans had not only seen moa, but had eaten them too.

In 1839, a shard of moa bone reached Richard Owen at the Hunterian Museum, where he was responsible for conserving the surgeon's collections and expanding them in the name of science. This fragment was to become pivotal in the career of the man who coined the word dinosaur, who commissioned the Crystal Palace monsters, and on whom Charles Dickens would draw for Mr Venus, the melancholy taxidermist and articulator of bones in *Our Mutual Friend*.

It was not a speedy process. After four years' deliberation, Owen decided that the bone belonged to a huge bird which he named *Dinornis* in a taxonomic echo of his terrible lizards (and, unbeknown to him, a nod to the future revelation that birds themselves were direct descendants of the dinosaurs): 'So far as my skill in interpreting an osseous fragment may be credited, I am willing to risk the reputation for it on the statement that there has existed, if there does not now exist, in New Zealand, a struthious bird nearly, if not quite, equal in size to the Ostrich, belonging to a heavier and more sluggish species.'

Forty years later, celebrated and often criticised for such leaps of faith (and indeed for failing to credit others' discoveries), the dome-headed professor was photographed alongside a giant moa skeleton at the British Museum, of which he was now the director, wearing a tattered gown which, along with his disconcertingly staring eyes, made him look rather like a moa himself.

Even before Owen 'discovered' the giant bird, tales of its apparent survival were emerging. In 1823, George Pauley claimed to have seen a huge bird when he was walking near a lake in Otago: 'I ran from it, and it ran from me.' Later, in 1850, engineers prospecting a new railway line to Canterbury saw two large birds, bigger than emus, on the hillside. And in 1878 a farmer described a moa in the same countryside, its unmistakable shape – although no living human was supposed to have seen one – silhouetted 'for fully ten minutes on the brow of the terrace, bending its long neck up and down exactly as the black swan does when disturbed'. Such stories, that moas might still be stalking their way through the wilderness, still persist, although little evidence exists to substantiate them, beyond blurred photographs, promises of plaster casts and rumours of giant nests in dead kauri trees.

As Bill and I look across to the ocean from Fyffe House, the sun is already sinking. Ahead of us, the Kaikoura Canyon plunges more than six miles down to the ocean bed, the tail-end of the great chasm that marks the meeting of the Pacific and Indo-Asian tectonic plates, an abyss into which whole volcanoes are dragged down by the inexorable movements of continental drift. Extensive biomasses have been identified in the canyon, one hundred times more than expected in such waters; massive accumulations of life, from burrowing sea cucumbers to spoonworms, testament to this meeting of cool and warm currents that encourages upwelling nutrients and the innumerable organisms which feed on them.

And down there also swim the largest cephalopods known to humans, and some unknown to us, too, perhaps. Giant squid with barbed tentacles, arms up to forty feet long, and eyes the size of basketballs to allow in what little remains of the light. Tennyson conjured up just such a creature,

> Far, far beneath in the abysmal sea,
> His ancient, dreamless, uninvaded sleep
> ...Then once by man and angels to be seen,
> In roaring he shall rise and on the surface die.

These real-life krakens present a terrifying vision; or at least they would, if anyone had ever seen them in their natural state. But they too, like many of these islands' animals, seem to be invisible.

The next morning I get up three hours before dawn, pausing briefly to marvel that each encounter with a cetacean seems to demand ever-earlier risings. I dress quickly, sling my roll bag over my shoulder, and leave quietly by the back door, stepping out into the night-morning, then trudge down the hill to the shore. The Southern Cross binds the sky, pointing to the South Pole.

I walk between yellow pools of street lights. The narrow park on the other side of the road, overlooking the shore, is decorated with an arbour of whale bone arches. The sea has stopped the roaring I could hear from my bedroom, a low rumbling that, in my restlessness, sounded like a perpetually-arriving train. I fight my imagination, which wants to fill my head with images of what might lie beyond the beach.

Reporting for duty in a fluorescent-lit room, I tug on my wetsuit. Half an hour later, I'm tipped off the back of the boat and into the still-black ocean, falling through the foaming surf churned up by the propellers.

Dark fins describe a wide circle a hundred yards away. It's difficult to see what's happening in the half-light – still more so since I'm hanging at sea level, waves slapping at my face. Frantically I tread water, trying to keep upright.

Suddenly, scything through my misty horizon, they're all around me. I'm in the middle of a super-pod of two hundred, probably many more, dusky dolphins.

I see their shapes, exquisitely airbrushed black and white and pearl-grey, swimming beneath me. Steadily, the fins begin to gather and steer towards me, more and more, till I'm in an eddying mass of swooping, diving cetaceans.

Everywhere I look there are dolphins; I'm encircled by them. They shoot from a single source like a shower of meteorites, their two-metre bodies zipping past, in and out of focus. I feel strangely calm. It's hard to believe this is real at all, as if the frenzy were happening somewhere else entirely.

But it's here, in front of my face. As I look down into the water, one animal slips into place at my side, then swims round and round, daring me to keep up. I spiral my body, only more aware of my ineptitude in their environment.

Dolphins are breaching right by me, turning somer-saults in the air. How about this? Can you do that? I reach out, instinctively; they easily evade me. That's not part of the game.

Dizzy and elated, I'm about to haul myself back on the boat when the skipper, Al, holds out his open palms, indicating I should stay where I am.

Is something wrong? He points over my head.

I look round and see dozens of dolphins heading straight at me, like a herd of buffalo. For a moment I think they're going to swim right into me. A ridiculous notion. They, like the whales, register my every move, my every dimension, both inside and out, my density, my temperature, what I am, and what I am not. A dolphin's sonar, which can fire off two thousand clicks a second, is able to discern something the thickness of a fingernail from thirty feet away. At the last minute the animals swerve aside, under my legs, by my side, past my head.

Many are having sex. With the males' two-kilo testes and penises that quick-release from genital slits, and females whose receptivity is advertised by plump flashing bellies, dolphins mate continually. Al says a single female may mate with three different males in five minutes, and will even mate with other species, producing dusky-common dolphin hybrids. As Caspar Henderson observes in his *Book of Barely Imagined Beings*, males will even insert their penises in the shells of turtles or the rear ends of sharks; females have been seen riding piggy-back, with one animal's dorsal fin in the other's genital slit.

Everything is turbulence. The water is alive with clicks, as if a current were being passed through it. I feel the sensual power of their bodies as they race past. But the space between us cannot be closed. Nothing passes in between. There is no connection. As abruptly as they came, they are gone. One dolphin takes two or three last spins around me. Then the waters fall quiet once more.

Back on the boat, I crouch, shivering on the prow, to watch the super-pod pass by. They've fed through the night and are content to play. One dolphin swims below the bow,

carrying a piece of seaweed in its beak. Māori navigators, *tohunga*, would appeal to dolphins for assistance in a storm: the *tohunga* would pluck a hair from his head and throw it to the *taniwha* or water spirits as a sign of their need for help – a tradition begun after they saw dolphins presenting seaweed as gifts to one another.

That afternoon, I board another boat on the far side of the Kaikoura Peninsula. The mountains that I'd ignored on my arrival, too close and too big to see, assert

themselves from the perspective of the sea, monumental mirrors of the underwater canyon. The animals take on a similar scale. The southern royal albatross, the *toroa*, with its wingspan of up to eleven feet, is, along with its wandering cousin, the largest seabird in the world. It is also one of the longest-lived: females have been found still breeding in their sixties. Gliding on enormous wings, one circles the boat, watching us through what Ishmael saw as 'inexpressible, strange eyes'.

A blue shark swims by, followed by a pair of blue penguins. Our boat tucks into the gentle swell as we come to a halt, and wait.

I'm on the upper deck, talking to the crew, when there is a commotion in the water ahead. Without any other warning, the huge blunt head of a sperm whale rises perpendicularly from the waves, with its mouth open. I can clearly see its massive canines. And caught between them is a three-foot-long kingfish.

Until now I thought that all sperm whales fed deep down; here was a whale eating at the surface. One of the crew points out the traces of the whale's sonar buzzes – much more powerful than the dolphins' clicks – creating localised circular patches, as if a ray gun had been trained through the water.

As the whale reappears, at full length this time, I can judge its bulk; the distance from its cantered nostril to the back of its skull is enough to indicate its size. Compared to the females I'd seen off the Azores, this bull is massive, at least fifty feet long and perhaps forty years old. Three decades ago it left its family, moving south in search of bigger prey, making itself more handsome and huge, and therefore more attractive to a partner. Now it has joined other males, both resident and transitory, to feed on the canyon's rich resources. There is no ignoring this animal. It is in its prime, regnant and supreme. But here too its kind are diminished, in number if not in size. Those resources may now be dwindling, and there are fewer and fewer sperm whales to watch, for all the frothed defiance of those that remain.

The whale dives, its flukes against the mountains, angling down into the canyon, all of a scale. Our captain passes me a pair of headphones which would look more at home on the New York subway, and I listen to the loud clicks below us. He smiles as he tells me, 'We call him Tiaki. It means guardian.'

8
The silent sea

Look across the beach from the sea, there is
what the mind's eye sees, romantic, classic,
savage but always uncontrollable.

GREG DENING, *Islands and Beaches,* 1968

To reach Kapiti Island one must, if not in possession of a driver's licence, board an early-morning train from Wellington. The line rumbles past colonial bungalows with verandahs and phormiums, New Zealand flax, staked by dead flower spikes, their blade-like leaves flapping in the ever-present wind. The day before, I'd stood at Island Bay to the south of the city, watching that wind bring with it one of the most violent storms seen in months.

I'd gone there in search of orca; a pod had been spotted in the harbour. Instead I was greeted with a tempest. The darkening clouds set the sand into sharp relief. One, then two rainbows arched over the horizon, joining the land and the sky. The intensity of colour and the falling air pressure had a hallucinogenic effect. Speeding towards us across the water, the sky was sucking up the sea, creating curls of feathery vapour that seemed about to turn into tornados. The full force of the tropical Pacific was meeting the chill of the Southern Ocean. It was like watching a weather forecast simultaneously slowed down and speeded up. The storm was rushing towards me. In an instant, the atmosphere became supercharged, an almost tangible mass.

I could practically feel the electricity crackling in the air. At the last moment, as the whirling wind rocked vehicles in the car park and fired perpendicular hail, catching up everything in its violent breath, I ran for cover. Waves which had rolled unobstructed over the ocean, gaining size and strength from immense gyres, were pummelling the rocks as if to tear them out of the sand. Nature was having a fit of hysteria, like an overactive child.

The next day, unless you read the newspaper headlines, you would have been forgiven for thinking nothing had happened. In the calm of the morning the train trundled on, leaving the suburbs behind for a series of beach settlements. As the sea began to reappear, in the distance, lying low, was Kapiti Island, its dark-forested outline just visible across the narrow strait.

New Zealand is one of the last places in the world to have been occupied by humans. In its sixty-five million years of isolation, it produced perhaps the greatest variety of bird species on any island, estimated at nearly two hundred. Like its whales, they too were a wonder and a resource to the Māori, who wove capes from their feathers, and threaded dead or living birds through their earlobes and kept them there till their wings stopped flapping and their bodies began to rot. When James Cook first arrived in 1769, he claimed that the birdsong, heard from the *Endeavour* at anchor in Queen Charlotte Sound, was almost deafening, as if a thousand finely tuned bells were ringing. But within a hundred years, animals introduced by the settlers were threatening that resonating abundance, and in 1897, in an early act of conservation to echo that of St Cuthbert on Inner Farne, Kapiti was declared a bird reserve. It has since witnessed the remorseless destruction, under the aegis of the Department of Conservation, of every mammal on the island, a violent means to a righteous end.

First they came for the domesticated animals, the cows, the pigs, the goats; all shot or slaughtered. Then they turned to the rabbits; trapped and dispatched. Most persistent were the bush-tailed possums, imported from Australia in the late nineteenth century to start a fur trade on the island; more than twenty thousand were destroyed. Then work turned to the rats. Soon enough, Kapiti was predator-free, leaving only its autochthonous avian population.

Human visitors are strictly regulated, and instructed to search their backpacks for any stray rodents. I check mine: no beady eyes down there, nor even a stray bit of fluff. A tractor tows our boat on its undercarriage, like a piece of artillery, over the grey sand and out into an uninviting sea. The wind in my face revives my hopes for the day. Every island, no matter how large or small, promises a story, a narrative of its own. 'Every living thing on an island has been a traveller,' wrote Greg Dening, the Australian Jesuit priest-turned-anthropologist. 'In crossing the beach every voyager has brought something old and made something new.'

Cook called it Entry Island. Its Māori title is scarcely less prosaic, meaning boundary – Kapiti marked the division between the *rohe* or territories of two *iwi*. But then, *Māori* itself means 'normal'. And though its names may be workaday, the island is not. Tui, stitchbirds, bell-birds sing and whistle in the forest. Tree ferns erupt from the dense undergrowth, turning the light itself green. It's like walking on the bottom of the sea.

After an hour's climb I break out of the gloom, onto the top of the mountain. As I sit to eat my picnic, a weka, sleek and brown, a cross between a turkey and an enormous starling, calmly walks up my legs in search of crumbs. I look over the trees to the sea far below, swirling around the island's rocky skirts.

I'm on my way down when I hear someone singing ahead. Who would disturb this forest, the kiwi sleep-ing in their burrows and the toutouwai pecking in the dirt? As I turn the corner, I realise the sound is coming from the Māori guide who'd accompanied us on the boat over. She's calling to a kokako, a rare wattlebird, invisible in the canopy above; and it seems to be calling back in return.

Back on the beach, I pick my way over bits of bleached driftwood, shattered *pāua* shells and cable-thick kelp, trepidatious yet determined to swim. Further down the strand stand two rusting try-pots. A fur seal slides into the sea and rolls onto its back, pressing its flippers together like a praying monk. These places smell of whale.

As I wait for the boat, a kaka peers at me from the lower branches of a tree, cocking its head in the questioning, almost mocking manner that parrots have. Fixing me with its eye, it declines the invitation to perch on my arm, and flies off into the canopy.

One afternoon – or it may have been an evening, or a morning, I don't know – in 1824, an eminent physician made an unusual house call in Liverpool. He was to attend a case of measles, a common enough disease. But his patient could hardly have been more unusual, since the entirety of his face, like much of his body, was tattooed, to the extent that from a distance his features appeared to be blue-black, inhuman. He looked more like a demon.

Even in the streets of a great port such as Liverpool, this man's appearance was remarkable. His name was Te Pehi Kupe, and in his homeland he was a celebrated warrior; yet here he was, tailored in Western dress, a dandy's cravat around his neck. It was clear to the doctor that there was a strong attachment between this exotic figure and the sea captain in whose house he was lodged. Something had happened to bind these two men together. Their visitor was intrigued – not least, perhaps, because he was an islander himself.

Thomas Stewart Traill was born in 1781 in Kirkwall, Orkney, far off the north Scottish coast, where he had grown up with whales and seals and seabirds; Orkney means 'seal islands' in Norse. Its neolithic houses, whose stones are still embedded in its turf, were built of whale bones and held sacred objects carved from whales' teeth. Like the Māori, their inhabitants had relied on the sea, rather than the land, and knew its animals well.

As a young man, Traill had studied in Edinburgh, famous for its Enlightenment spirit, and he now practised medicine in Liverpool. But he was no mere physician. In the vital issues and debates that concerned educated men of the age, Dr Traill was one of the inner circle that Sir William Roscoe, Member of Parliament, banker, historian, penal reformer and abolitionist, gathered around himself. Liverpool, from where my own ancestors would leave for America or arrive from Ireland, had been 'the chief seat of the odious traffic' of slavery, the starting and finishing point in the terrible triangle that bound Africa to the Caribbean and Britain. It was a horror in which one of Roscoe and Traill's contemporaries saw an equivalence with the way we treated other species. Jeremy Bentham accepted that man might kill them for food, 'But is there any reason why we should be suffered to torment them? Not any that I can see. Are there any why we should *not* be suffered to torment them? Yes, several.'

The philosopher and reformer was certainly eccentric in his ideas. In 1824, the same year that Traill met Kupe, Bentham had become fascinated by the Māori method of preserving human heads, and ordered that his own should be displayed after his death as an Auto-Icon, along with his soft body parts, labelled in decanters in the same cabinet. As the inventor of the Panopticon, Bentham had addressed the keeping of humans; now he extended his gaze to animals, in this age of menageries. He might have been writing with the fate of Chunee in mind; or indeed any other hunted or confined creature, whose treatment, as Blake had written, 'puts all Heaven in a rage':

> The day has been, I grieve to say in many places it is not yet past, in which the greater part of the species, under the denomination of slaves, have been treated by the law exactly upon the same footing, as, in England for example, the inferior races of animals are still. It may one day come to be recognized that the number of the legs, the villosity of the skin, or the termination of the *os sacrum* are reasons equally insufficient for abandoning a sensitive being to the same fate... the question is not, Can they *reason*?, nor Can they *talk*? but, Can they *suffer*?

Sir William Roscoe's own views on animal liberation may go unrecorded, but like Bentham, he did not limit his energies to opposing the evils of slavery. As with so many men of his position and taste, he was an inveterate collector – not least of people. He counted among his friends

Horace Walpole and Henri Fuseli, and he corresponded with notable Americans such as Thomas Jefferson and Washington Irving. Evidently, he was keen to establish relations with the new republic. In May 1818 he had invited Allan Melvill, a Manhattan trader in fine goods, to his house; thirty years later, Melvill's son Herman would visit the same port, having added an 'e' to the family name.

Liverpool, like Southampton, was a great gateway, open to the world. It seemed to summon such an eclectic and oddly connected cast, among them another of Traill's friends: William Scoresby, latterly of Whitby, at that point resident in Liverpool, where he preached in a dockside Floating Chapel. Scoresby had been a champion whaler, like his father; but he was also a scientist and vicar, professions which he saw as quite compatible. 'They surely will

DELPHIN
Le

not deem it intrusive,' the Reverend Scoresby informed his readers, from the pulpit of his *Account of the Arctic Regions* (a book which Melville would plunder shamelessly for his own) 'to be reminded that the most important preparation for such undertakings, as well as for the whole of life, is to surrender the heart to that Saviour who has died to redeem his servants from guilt and ruin.' It was a time in which faith and exploitation were necessarily intertwined. 'Reader, do you understand, and have you accepted, this gracious message?'

Such was Scoresby's admiration for Traill that he named an Arctic island for his Orcadian friend. The gesture was an acknowledgement of Traill's own scientific cataloguing, not least as the first man to study the pilot whale, as Scoresby noted: '*Delphinus deductor*, defined by

DUCTOR.
eet

Dr. Traill... This kind of dolphin sometimes appears in large herds off the Orkney, Shetland, and Faroe islands. The main body of the herd follows the leading whales, and from this property the animal is called in Shetland the ca'ing whale, and by Dr. Traill the deductor... in modern times extensive slaughters have taken place on the shores of the British and other northern islands.'

From slavery and science to mesmerism and whaling, these men were passionately interested in the issues of their age. And what more extraordinary case history than this tattooed figure from the other side of the world, whose story was as intriguing as any taxonomic study of a blackfish? Perhaps Traill – who would later assist John James Audubon in publishing his *Birds of America* – saw Kupe as an exotic coloured book plate. Yet in the role of doctor, he also understood that his new patient might be about to die. The Māori had been the subject of an experiment, having been inoculated with measles by a surgeon – possibly following the example of Edward Jenner, John Hunter's pupil – and was now seriously ill.

The measles morbilliviruses infected newly discovered islands with a ferocity beyond even venereal disease, halving the Māori population within two generations of James Cook's first contact. Dr Traill used his lancet to blister the disease, an archaic and largely useless technique, widely adopted in lunatic asylums; Kupe's recovery probably had more to do with his strong constitution and the ministrations of his friend, the sea captain Richard Reynolds. Fascinated by the Māori, Traill invited him to his home, with a view to finding out more about him and his countrymen. For Kupe himself, this too was an experiment and an adventure. He also had a mission: something to gain and, perhaps, much to lose.

When George Craik, a fellow Scot and noted writer, was working on his book *The New Zealanders*, he turned to his friend Dr Traill for information. As a contributor to the wonderfully named Society for the Diffusion of Useful Knowledge, Craik was unabashed by his lack of first-hand acquaintance with his subject; he knew how to pique the curiosity of his audience. 'We are about to introduce to our readers a highly interesting native of New Zealand, who has recently visited our shores, but of whom, we believe, no account has yet been given to the public.' There was more than a hint of the gothic to his account of the alien; of Romantic imaginings, of Walpole's fantasies and Mary Shelley's science fiction. It is why, perhaps, Melville (whose book was influenced by his reading of *Frankenstein*) would adopt the same figure as an evocation of the other in *Moby-Dick*, incarnate in the persona of Queequeg; the dark obverse to the apparent security of Western civilisation. Those cryptic marks on Kupe's face evoked another world.

But Mr Craik was not concerned with any inkling of metaphysics. 'Of all the people constituting the great Polynesian family,' he noted, 'the New Zealanders have, at least of late years, attracted the largest portion of public attention... They present a striking contrast to the timid and luxurious Otaheitans, and the miserable outcasts of Australia.' Masculine, independent, hierarchical and resistant – as opposed to the apparently complaisant, property-disdaining Aborigines – the Māori were deemed worthy opponents, like the Zulus of South Africa, warriors fit to fight the Empire. 'From the days of their first intercourse with Europeans they gave blow for blow. They did not stand still to be slaughtered, like the Peruvians by the Spaniards; but they tried the strength of the club against the flash of the musket.'

Nor were the indigenous people of the South Seas strangers to British streets. The first Polynesian, Omai, had been brought back on the *Adventure* in 1774, to be adopted as Joseph Banks's ward and commemorated in a portrait by Joshua Reynolds, arranged in stately robes. The first Māori to visit Britain was Moehanga, brought from the Bay of Islands by John Savage in 1807. Arriving in London among the forest of ships' masts on the Thames, he feared he might be lost, as he might lose his way in a kauri wood. Church steeples awed him. Coming from an island where a cloak of flax or bark might take months to make, and assumed a sacred status in the process, London's unholy consumerism was shocking to him, with its shops full of clothes and houses stacked with furniture. In turn, the warrior aroused amazement, for all that he had exchanged his feather headdress for a silk hat. 'It was extremely inconvenient to take Moehanga to public exhibitions, or even to walk with him in the streets, on account of John Bull's curiosity,' complained Mr Savage. And although the Māori expressed due incredulity at St Paul's vast dome, when he met a missionary in New Zealand many years later, what he remembered most of

London was its plumbing, and 'how the water was conveyed by pipes into the different houses'.

By then, Moehanga's visit had been eclipsed by the celebrated chief Hongi Hika, who had arrived in England in 1820, along with his young warrior nephew, Waikato. They became the cynosure of high society, although Hika himself was more drawn to the wild beasts in the Tower of London's menagerie, particularly the elephant. He was unfazed by his audience with George IV, declaring, 'There is only one king in England, there shall be only one king in New Zealand.' In honour of such ambitions, the British monarch gave his visitor a suit of armour, a somewhat impractical present, although it was claimed that in one battle his helmet protected the chief from a bullet.

When the two Māori returned home, they promptly sold their gifts for guns and ammunition, weapons which allowed them to wage battles whose violence and subsequent cannibalism were so shocking that Waikato admitted he could not eat anything for four days. Nor would Hika's armour protect him. During a skirmish he was shot through the chest, leaving a wound that took a year to kill him; the chief would invite his warriors to listen to the wind whistling through his lungs, and witnesses claimed they could see through his body.

Such were the precedents for Dr Traill's patient, whose presence in Liverpool was not as innocent as it seemed.

On 26 February 1824, Richard Reynolds, captain of the merchant ship *Urania*, owned by the trading company of Stainforth and Gosling, was sailing in the Cook Strait when a formidable flotilla hove into view from Kapiti: three war canoes, loaded with eighty warriors.

Reynolds and his crew prepared for imminent attack – and as a South Seas trader, *Urania* would have been well-armed. But they could not have expected what happened next. The largest of the canoes, with its tall prow, drew towards *Urania*'s bow, and a man – evidently the leader – stood up. In broken but clear English, he demanded to be allowed to board the ship. Reynolds declined, but as he could see no weapons in the canoe, he allowed it to come nearer.

Was he inviting what happened next? The company that employed him was about to go bust; perhaps trade for *Urania* wasn't going so well; perhaps this foreign intervention was a welcome diversion. Or maybe there was something more, something unspoken. Certainly, Reynolds had some deeper knowledge of New Zealand, since he appeared to be fluent in the Māori tongue.

As the two vessels came to within touching distance, the gap between worlds and over oceans was broached. Kupe – I imagine his powerful legs bending, his muscular arms reaching out – sprang from his canoe and landed on foreign territory. It was a leap of faith.

Once aboard, Kupe turned to his war canoes and ordered them to back off – a gesture of conciliation. He made signs as to what he wanted – guns – and was denied. But he had saved his key phrase, words he'd learned and understood, for his final, audacious demand: 'Go Europe,' he said, 'see King Georgy.'

Reynolds had had enough of this pantomime. He was neither an arms dealer nor the captain of a passenger ship. He ordered three sailors to throw Kupe overboard. As they tried to do so, the warrior threw himself on the deck, grabbing a pair of ring-bolts so powerfully that it was impossible to pull him away 'without such violence as the humanity of Captain Reynolds would not permit'. Kupe read the situation correctly. He shouted to his canoes to turn back. He was on his way to Europe.

Reynolds tried to put Kupe ashore at the next opportunity, but the wind was against him. Giving up, at least for the moment, the Englishman, exhibiting his good manners, decided to make his uninvited guest comfortable. He offered the chief a bunk in his own cabin, in recognition of his status. Kupe stayed on board as *Urania* sailed across the Pacific to South America, and by the time they reached Lima he and the captain were on the best of terms – a friendship consolidated in a dramatic incident at Montevideo when Reynolds fell overboard. Kupe jumped in and caught the captain as he was about to sink. Holding him tight above the waves, the warrior swam with Reynolds until the two men could be rescued. From one leap to another, the intimacy was sealed between Māori and Englishman. Like Truganini saving George Robinson from the river Arthur, Kupe's plunge into the South Atlantic was a reversal of the usual roles of Westerner and islander, and all the more powerful for that.

When he read it, as he must have done, twenty years later, this story had a powerful effect on Melville. Queequeg, a tattooed prince and hawker of human heads,

321

is the most memorable figure in *Moby-Dick*, clearly based on a Māori warrior; the first Pacific islander in Western fiction. Like Kupe, he too saves a white man from drowning, diving from the deck of the *Pequod*. 'Stripped to the waist, [he] darted from the side with a long living arc of a leap. For three minutes or more he was seen swimming like a dog, throwing his long arms straight out before him, and by turns revealing his brawny shoulders through the freezing foam.'

Queequeg hails from 'an island far away to the West and South. It is not down in any map; true places never are.' He is insular in the way the rest of the crew are: 'They were nearly all Islanders in the *Pequod*,' says Ishmael, '*Isolatoes* too, I call such, not acknowledging the common continent of men, but each *Isolato* living on a separate continent of his own.' Unutterably other yet honourable even in his inky guise, Queequeg is an island in himself. Like Kupe, he is mysterious and brutal, his body a contour map of the unknown, carved by the bone of a whale. Both men are comet-like, almost extraterrestrial. Their miraculous appearances seem to foretell the future: in Queequeg's case, the fate of the *Pequod* and her misguided crew; in Kupe's case, the fate of his warring island nation. When he witnesses his crewmate's heroic act, Ishmael is beyond admiration. 'From that hour I clove to Queequeg like a barnacle; yes, till poor Queequeg took his last long dive.' So too Kupe and Reynolds were bound to one another.

Te Pehi Kupe was born at Kawhia on the North Island around 1795, making him about thirty years old when he met Captain Reynolds and Dr Traill. He had two wives, one son and five daughters. In 1821 he had joined his nephew, Te Rauparaha – destined to become New Zealand's most famous chief – in raiding the fortified *pas* of rival *iwi*. They took Kapiti in a vicious battle, only after four of Kupe's own children were killed. The island became their stronghold, its warriors armed with muskets acquired from the whalers who continued to arrive in their thousands to kill whales even as these bloody land wars were in progress. And it was in pursuit of that internecine conflict, and perhaps to steal a march on his nephew, that Kupe left Kapiti, for England.

Arriving in Liverpool, with a Māori in tow, Reynolds found himself out of work, his employers having declared bankruptcy. No one would have blamed him if he'd given Kupe a few shillings and left him at Prince's Dock to fend for himself; or if he'd done as precedent suggested and exhibited his friend for a fee, much as one might do with a stranded whale. That Reynolds did neither was a measure of the two men's attachment – one which Dr Traill witnessed for himself. On his visits, the doctor observed that Kupe became upset if parted from Reynolds for more than an hour; his loyalty was absolute, and he even moved the captain's luggage into his own room, 'for fear his friend and protector should be carried away from him'.

On examination, the patient was found to be 'yet in the vigour of life... His face was intelligent and pleasing, though so much tattooed that scarcely any part of its original colour remained visible. Indeed, every part of his body was plentifully covered with these marks.' Like Queequeg, whose skin resembled a quilted bedcover it was so patterned ('Good heavens! what a sight! Such a face! It was of a dark, purplised, yellow color, here and there stuck over with large, blackish looking squares'), Kupe exuded a physical appeal in Traill's admiring account: 'His finely muscular arms, in particular, were furrowed by a great many single black lines; and these, he said, denoted the number of the wounds he had received in battle.' And like Queequeg, too, Kupe's usually equable temper could occasionally flare. When a sailor on *Urania* had insulted him, 'he rushed upon the man, seized him by the neck and the waistband of the trowsers, and after holding him for some moments above his head, dashed him on the deck with great violence'. The scene recurs in *Moby-Dick*, when Queequeg catches a 'young sapling' mimicking him behind his back. He promptly tosses him in the air like a caber. Reprimanded by the captain for nearly killing the miscreant, the warrior prince replies, laconically, 'Ah! him bery small-e fish-e; Queequeg no kill-e so small-e fish-e; Queequeg kill-e big whale!'

Dr Traill and Kupe went riding together, a remarkable enough image. For his part, it was claimed that the first time he saw men on horseback, Kupe appeared astounded, like the Aztecs when they first encountered the conquistadors; although, aware of his place in the drama, perhaps the visitor was performing to his role, too. In many of his reported reactions to English manners and customs, it seems as if the Māori was trying to please his hosts with his 'savage ways'. When he found himself surrounded by immense crowds in the streets, Kupe acknowledged them by touching his hat and shaking their hands. He had become a personality, and his likeness was recorded by an artist, John Sylvester. The process fascinated the sitter, who insisted that his tattoos should be accurately copied. His *moko* was his identity, especially the markings over the upper part of his nose, which represented his name (although, ironically, that name was Westernised as Tupai Cupai). Kupe drew the *mokos* of his brother and son, and pointed out the differences between them. He knew every line on his body from memory, although the only mirrors in his native land were reflections in gourds of water.

Portrait of Tupai Cupa.

On his rides in the Lancashire countryside, Kupe was fascinated by agriculture and blacksmiths. One day Traill took him to see a review of a regiment of dragoons, 'a spectacle of course altogether to his taste', as George Craik reported. 'The gay appearance of the troops – their evolutions in making a charge – and the command which the men exercised over their horses, – all drew from him the warmest expressions of wonder and delight.' It was his cue to reiterate his true agenda. He asked if the king had many more such warriors; on being informed that he did, Kupe replied, 'Why then he not give Tupai musketry and swordy?' and offered to pay in spars and flax.

This was no naïve barter. The spars of which Kupe spoke came from the kauris that, he told Traill, grew down to the shores of Kapiti as if full to bursting, and were valued for ship's masts; plentiful spikes of flax furnished the raw material for sailcloth. Kupe knew just what he was offering to a naval empire (whose ambitions differed from his own only in scale and materiel). He even employed sentiment to persuade his friend. In a poignant scene, Kupe met Traill's four-year-old son. Taking the boy onto his lap, he kissed him and began to weep, telling Traill that his own son had been the same age when he was killed and eaten, his eyes scooped out and devoured.

Despite his wiles, Kupe failed to convince his hosts that it was a good idea to hand over the guns he wanted to wreak his *utu*, revenge. On 6 October 1825 he sailed from England, at the country's expense, aboard the *Thames*. With him he took various agricultural implements donated by the government; doubtless it was hoped that these would encourage his people towards more pacific pursuits. But like Hongi Hika before him, Kupe quickly traded the tools and his Western clothing for guns as soon as he reached Sydney.

Back in Kapiti, Kupe plunged into apocalyptic, intertribal war. And like Hika, his end came almost inconsequentially, in an argument over a valuable piece of nephrite. 'Why do you with the crooked tattoo, resist my wishes?' he is said to have told a rival warrior, '– you whose nose will shortly be cut off with a hatchet.' Appearances were important, even unto death. Soon after this, Kupe's forces were overwhelmed. As he resisted, his final words were ambivalent – 'Don't give it to the god, but to the Kaka-kura' – but his fate was not. His flesh was cooked and eaten, and his warrior bones, which his country would not have, were made into fish hooks to be dangled into the ocean. Like Cook's, his was a symbolic death, an island sacrifice to the sea; an end, and a beginning.

The early-morning bus leaves from under the lee of the cathedral, which a year later would lie in ruins, and winds its way through the steep-sided hills of the Banks Peninsula to the sea. Next to me, a dreadlocked girl spends most of the journey deleting images from her digital camera, bleep-bleep-bleep, one by one, providing an electronic soundtrack to our descent to Akaroa – 'long harbour' – set inside a narrow rocky inlet.

The Banks Peninsula was named by Cook in honour of his patron; or rather, he called it Banks Island, since that's what his initial survey indicated it was. Then it was covered in trees, but these were soon shipped back to Europe for masts, leaving its hillsides bare. That sense of destruction was reflected by bloody warfare: it was here that hundreds of Māori died in a brutal attack by Te Rauparaha, abetted by a British captain, John Stewart, in 1830.

By that point, this peninsula had become a profitable shore-whaling centre for the Europeans who first settled here. Around its headlands, in remote coves, lie the crumbling remains of whaling stations, lingering evidence of early deals done for this place's resources. In 1838 Akaroa nearly became a French colony under another captain, Jean-François Langlois, sailing on the Le Havre whaler *Cachalot*. He bought most of the peninsula's land from the local Māori, who received, in part payment, two cloaks, six pairs of trousers, two shirts, twelve hats, two pairs of shoes, some pistols and a few axes. At one point Langlois suggested the peninsula as a penal colony for his motherland. He set sail from France with his first group of free settlers, only to discover that by the time he arrived in August 1840, the Māori had resold the land to the British, who had established their control of New Zealand with the signing of the Treaty of Waitangi, and had planted the Union Jack in Akaroa.

Akaroa has not forgotten that it was nearly French. The Tricolore still flies in the centre of town; quaint gabled houses bear such names as La Belle Villa; you all but expect to see onion-sellers cycling down the streets. It is a peaceful, almost stage-set place, not unlike Freshwater Bay; a sense of sleepiness, if not complacency, masking its history. But it is not its history that has brought me to this last near-island. It is an animal named after an Englishman, and now one of the rarest of its kind.

I'm lying in my wetsuit on the wooden pier in the sun, almost falling asleep, when I'm called. The boat is ready to leave. I clamber aboard, and we sail out towards the open ocean. The forecast is for high seas. I chat to the skipper, a New Zealander named Ian, and Alan, his first mate, a young Welshman. The water here is turbid – the perfect conditions for the Hector's dolphin, named after Sir James Hector, Victorian director of Wellington's Colonial Museum. They are the smallest of all dolphins; calves are barely bigger than rugby balls, and adults hardly more than a metre long.

Unlike their acrobatic dusky cousins, *Cephalorhynchus hectori* do not immediately announce their presence with backflips and breaches. Their subtle circular fins slice through the waves before you have the chance to establish what they are. Small and fleet, they prefer the sanctuary of cloudy, coastal water as protection against predators, principally sharks. But those same waters make them vulnerable to human actions, subject to the pressures of a maritime nation which owns more boats per head than any other on earth. Stressed by noise, poisoned by chemical and agricultural pollution, and drowning in undiscriminating gillnets, as Dr Barbara Maas observes, these animals are dying more quickly than they breed. The North Island subspecies, known as Maui's dolphin, constitutes just fifty-five individuals; as the most endangered of any cetacean, they are unable to sustain any more losses. They are on the brink of disappearance, here, at the end of the world, about to go the same way as the thylacine and the moa.

It is almost impossible to imagine that these animals might not survive the century; that I could outlive them. While the deeper waters around New Zealand conceal species of beaked whales that have never been seen alive, the sheer energy of these diminutive cetaceans so close to shore seems to defy any notion of extinction. There's something cartoon-like about them; not just in the way they move but in their rounded dorsals, like the handles on Continental coffee cups, the kind you can't put your finger through; I might lean over and pick one up for a closer look.

Togged-up and raring to go, like a greyhound at the gate, I perch on the diving platform at the back of the boat, blowing through my snorkel and slapping the water with my fins. 'You've done this before, haven't you?' says Alan.

The skipper shuts off the engine and I leap in. It's terrifically difficult to tread water. The grey-white bodies begin to circle around me; it's like being surrounded by sheep dogs. Then they begin leaping into the air, showing off their beautifully marked bodies, like little aeroplanes. Their black-and-white masked faces and striped bellies might as well be *moko*. But my body declines to act like a dolphin, and there's as much seawater up my nose as there threatens to be in my lungs. I feel like a circus animal myself; and although the dolphins come to me of their own accord, I can't help feeling they'd be better left alone. Exhausted, I haul myself up over the boat's side. Sometimes it's better to watch than to take part.

Back in the guesthouse, in a tiny room made smaller by lace and chintz, I take two painkillers and fall asleep, to the memory of the rocking waves. I wake at dusk, still woozy with the sea. I wander through the empty town, closed-up and out of season. I look in through the windows of the only restaurant still open and see a handful of diners bent over their meals.

I buy a bag of chips from the nearby takeaway and retreat to a bench overlooking the harbour. Afterwards I walk to the end of the pier, looking out to sea. I'm weary of being away, of shabby oil-stained shorts and basin-washed T-shirts, of being removed from family and friends by half a dozen time zones. I feel homeless, rather than homesick, faced with the familiar strangeness of this place, so orderly like home, yet on the edge of utter wilderness. More than anything, I feel abandoned, as I always have.

My notebook sits on my bedside cabinet. Everything is invested in its pages: the postcards and dried leaves and ticket stubs I've stuck in it, the slivers of whale skin, the sketches of unknown places and animals. In the absence of anything else, it is my home, my life spiral-bound between black card, the anchor I let down.

Maybe I'll never get back. In my dreams I sink into invisible magnetic fields and invisible suffering; my head is filled with migrations and invasions, travellers and victims. All the while, giant albatrosses glide and great whales dive into abyssal canyons, and Hector's tiny dolphins play hide and seek in the water.

It's time to go home.

9
The sea in me

Time has never existed, and never will; it is a
purely artificial arrangement. It is eternity now,
it always was eternity, and always will be.

RICHARD JEFFERIES,
The Story of My Heart, 1883

There could hardly be a more common bird, yet you could travel around half the world and never hear anything so beautiful as a blackbird in a suburban garden. Their big eyes sense the slight slip from darkness to the semblance of light before all other garden birds; only robins can rival them in this keen awareness. I listen to the first notes of the first song, a lone voice in the dark, joined by another, then another, until they form a circle of sound. From dawn to dusk they rise and fall, fit and start, from roofs to trees, announcing their allure. Their songs are asymmetrical, apparently random; phrases are thrown out to be echoed by rivals, in the way humpback whales take up that year's song and repeat it through the oceans. As the philosopher and musician David Rothenberg showed me, when you speed up the song of a humpback, it sounds very much like birdsong, with the same 'sustained whistles, rhythmic chirps, and noisy *brawphs*'.

Each sequence is its own narrative, precisely measured out. Blackbirds have the ability to sound both ridiculous and sublime at the same time, with their querying intonation ending in an upnote, like a teen's mallspeak – duh-duh-*duh*?; or duh-duh-*lu*, duh-duh-*lu*! But theirs is a serious intent, bent on preventing any incursion into their fiefdom, as well as sounding sexy to a potential mate. They'll fly just a few feet off the ground, to evade predators from above – although a habit which made sense when their only enemies were raptors is less useful now that their low flight-paths take them directly into potentially lethal traffic; it amazes me, as yet another black streak almost zooms through my bike wheels, that they don't sustain more casualties. They must retain a race memory of when all this was only heathland. A blackbird defends its territory all its life; some may live for up to twenty years. The same bird bobs and bows and runs across my garage roof year after year, looking up at me in turn.

How can such a grey, wet day be so beautiful? After days of rain I ride out at dawn, taking my chance during a brief interlude of dryness. There's nothing to focus on, just cloud. Under such skies, anything is a gain. It's May Day. The rain intensifies the smell of the morning. The woods through which the road runs lean over and meet tree-to-tree, negating the tarmac below. At the beach, the water is flat calm. The world has opened up again.

A new shape appears high over the shore; the slender wings of a swallow, zigzagging its way from the sea to the trees, thousands of miles from sub-Saharan Africa. Later, I'll watch them from water level as they swoop within inches of my head, so close that I can see every detail: blue-black backs as iridescent as a mineral, pure white bellies and rosy chins. To the Romans, the swallow represented the household gods because it nested in the eaves; it was unlucky to kill one. But its name comes from Scandinavia, whose early Christians believed that it had flown over the Crucifixion crying *Svala! Svala!* – Console! Console!, and called it *svalow* in tribute to its piety.

The birds' annual disappearance was a source of mystery. Some said they flew to the moon, or even changed species. As late as the sixteenth century it was believed that they hibernated in the water, from where fishermen could cast their nets and pull out swallows, 'huddled against each other, beak to beak, wing to wing, foot to foot...among the reeds'.

When Gilbert White watched great flocks getting ready to leave, the sight touched him 'with a secret delight, mixed with some degree of mortification', since no one yet knew where they went. 'If there are any animals with no memory they may be happy,' wrote his descendant T.H. White, 'but even swallows remember last year's nest.' To Ted Hughes the swallow was 'a whiplash swimmer', 'a fish of the air', 'the barbed harpoon'. For sailors they were bluebirds, heralds of home; a swallow tattoo would ensure its bearer's safe return to dry land, in the same way that an anchor symbolised hope. But my artist friend Angela, who saw her first bluebird weeks ago in Cornwall, tells me that a tattooed dagger through a swallow's heart is the sign of a lost loved one.

I wash off the night in the water, my scrapes and aches numbed by the sea. My bones have become boughs, all scarred knees and gnarled knuckles. None of us are the same person we once were, since the human body entirely replaces itself every seven years; there have been at least six different mes.

A few days ago a pair of common terns appeared. They were once called sea swallows for their forked tails – hence their name, *Sterna hirundo*, from the Old English *stearn*, and the Latin *hirundo*, swallow – but they also resemble stripped-down, go-faster gulls, all cackling cries and electrical energy. They scan and dive, feeding furiously to replace calories lost on their own long flight from Africa. They look as though they'd barely make it over the Solent, but the Arctic tern holds the world record for bird migration. One was found to have flown from Finland to Western Australia, a journey of fourteen thousand miles.

The winter birds have all flown. One morning in March, we'd descended on the shore with cannon nets in an attempt to catch brent geese before they left. Our leader marshalled his troops with military ease, taking the sightlines of the metal tubes embedded in the mud, ready to fire over our unsuspecting prey. We stood, and waited. A passing dog cocked its leg on Peter's bag. Suddenly, a bang and a puff of smoke, and we ran to retrieve the netted birds, as if to drag them out of the water. I feared for one which had its head in a shallow pool. But in all eight brent were bagged in sacks – along with a single oystercatcher.

The nine wriggling bundles of hessian lay on the shingle awaiting processing, the more ambitious attempting to escape, scrabbling inside as if in an avian sack race. The oystercatcher was first out. Ruth held the now-ringed bird, showing me how its head and eyes remained focused on the ground while she moved its body like a compass, its bright red bill always pointing to its food. Then she handed it to me, and I let it loose. Its white-flashed wings vanished in an instant. We set to ringing the geese.

I felt the faint oiliness of the birds in my hands; saw their neat black heads, the subtle white collars around their necks, their serrated bills trying to peck my fingers. Up close, they were much smaller than I'd expected, more duck-sized than goose, and yet so fearless and intrepid: fine, proud, wild creatures, Arctic emissaries to this suburban shore. We set them free together, and soon they too were gone, their burnt-grey shapes fading into the sky.

Back home, I pluck up the courage to clear my mother's room. The bed in which she lay, and from which she left to go to the hospital for a routine operation and never returned, has not been disturbed for six years. I left it, reassured by the smell, as if the room, like the house, were still full of her. Now I pull back the bedclothes, heartlessly. Twenty-four hours later, it's all gone. The room is bare and empty. From the windows where the curtains always seemed to be drawn, I peel the sticky plastic film that imitated stained glass in the fanlights. As it comes away, with a final yank that sends me flying backwards off the stool on which I'm balanced, the light pours in.

All the things I imagined as a child, all the things I feared; they're not at the end of the world, and they're not here, either. I close my notebook and put it on the shelf, along with all the others.

There's no such place as home. And we live there, you and me.

Acknowledgements

I would like to thank my brothers Laurence and Stephen, and my sisters Christina and Katherine, and their partners; Harriet, Jacob, Lydia, Max, Oliver and Cyrus for their comments; and Mark Ashurst, for his steadfast support. For their encouragement and advice: John Waters, Michael Bracewell, Neil Tennant, Hugo Vickers, Ruth Wilson, Clare Goddard; Adam Low and Martin Rosenbaum, for our antipodean adventures together; Andrew Sutton and Rachel Collingwood for their excellent company; and Angela Cockayne, for her constant inspiration.

My editor, Nicholas Pearson for his unfailing enthusiasm, and Olly Rowse for his Anglo-Saxon attitudes; Robert Lacey, Julian Humphries, Terence Caven and Patrick Hargadon for their work on the text, design and publicity. A special thanks to Joe Lyward for his beautiful drawings. My sterling agent, Gillon Aitken, and his colleagues Anna Stein, Sally Riley, Imogen Pelham, Leah Middleton; and my Spanish publishers, Claudia and Joan at Atico do los Libros, Barcelona. I would also like to acknowledge, with gratitude, the support of the Leverhulme Trust as Artist-in-Residence, the Marine Institute, Plymouth University.

United Kingdom: Ken Collins and Jenny Mallinson and the National Oceanographic Centre, Southampton; Peter Wilson; Jolyon Chesworth and the Hampshire and Isle of Wight Wildlife Trust; Peter Potts; Lydia Fulleylove and the writers of HMP Albany;

David Chunn; Mary Hallett; Fr Claro Conde; Tobie Charlton at Bike Guy; Angela Barrett; Hilary Franklin; Barry Udall, Nicholas Moore; Cynthia Walsh, Vanessa Williams-Grey, Mark Simmonds, Whale and Dolphin Conservation; Mark Carwardine; Dylan Walker and Ian Rowlands, Planet Whale; Dr Malgosia Nowak-Kemp, Oxford University Museum of Natural History; Dr Stephen Sleightholme; Anthony Caleshu, Sarah Chapman and Peninsula Arts, Plymouth University; Martin Attrill, Simon Ingram and the Marine Institute, Plymouth University; Ruth Leeney; Rob Deaville and Matt Perkins, Cetacean Strandings Investigation Unit, Zoological Society of London; Richard Sabin and Jon Ablett, Natural History Museum; Phoebe Harkins, Wellcome Library; Anthony Wall, Adam Nicolson, Tim Dee, Viktor Wynd, Mat Humphrey, Paul Bonaventura, Jeremy Millar, Brian Dillon, Alison Turnbull, David Gray, Peter Doig, Keith Collins, George Shaw, Marc Riley, Richard Hawley, Paul Ballantyne, Michael Holden.

The Azores: Serge Viallelle, João Quaresma and Espaço Talassa; Malcolm and Dorothy Clarke; Karin Hartman, Helena Cepêda; Sara Vanessa Santos; Rui Prieto; Secretariado da Direcção Regional dos Assuntos do Mar.

Sri Lanka: Geoffrey Dobbs, Shyam Selvadurai; Hiran Cooray, Anoma Alagiyawadu and Riaz Cader, Jetwing; Sanjika Perera, Sri Lankan Tourist Board; Dr Chris Nonis; Asha de Vos, Anouk Ilangakoon.

Australia: Cassandra Pybus, Iain McCalman, Andrew Darby, Hamish Maxwell-Stewart, Robert Dessaix, Dr Tony Brown, Thomas Keneally, Michelle de Kretser; Robert Pennicott, Tim Cunningham and Bruny Island Cruises; Ericka Bowd, 43 Degrees, Adventure Bay; Alan Smith, State Library of South Australia; Suzanne Miller, South Australian Museum; Dr Sandy Hume.

New Zealand: Anton Van Helden, Museum of New Zealand Te Papa Tongarewa, Wellington; Roger and Lisa Payne; Laura Kroetsch, Sandra Noakes; Mike Donoghue, Annie Wheeler, Bryan Jensen, Derek Cox and the Department of Conservation; Kauahi Ngapora and Lisa Bond, Whale Watch Kaikoura; Jo Thompson,

Alastair Judkins and Encounter Kaikoura; Manuel C. Fernandez; Ian Fitzwater, Nikau Lodge, Kaikoura; Black Cat Cruises, Akaroa; Bill Morris, Gregory Rood, Witi Ihimaera.

United States and Canada: Pat de Groot, for providing the sea outside; Dennis Minsky for the birds and whales; Mary Oliver, for her poetry; and Mary Martin, for the same; Dr Charles 'Stormy' Mayo, Josiah Mayo; Dr John Gullett; Captains Todd Motta and Mark Delumba, Dolphin Whalewatch, Provincetown; Dr Carole Carlson, Tim Woodman, Jo Hay, Meribeth Ratzel; Skott Daltonic, Provincetown Whales; Iain Kerr, Scott McVay, Phillip Clapham; Dr John Wise, University of Southern Maine; Alex Carleton, David Rothenberg; Dr Paul Pearson, Thomas Merton Center, Louisville, Kentucky; Hal Whitehead, Dalhousie University, Nova Scotia.

Philip Hoare
Southampton, March 2013

Text and Image Credits

346

Sources

For more detailed notes, please go to
http://www.philiphoare.co.uk/2013/03/27the-sea-inside-
source-notes/

1. The suburban sea

Richard Hamer, *A Choice of Anglo-Saxon Verse*, Faber 1970

'Fawley Refinery, 1953', www.britishpathe.com; 'Fawley Refinery'
www.exxonmobil.co.uk

'Methodology for the measurement of impingement', *British
Energy Marine & Estuarine Studies, Scientific Advisory Report
Series 2010*, No 006, Ed 2

'Spike Island & The Vagrancy Act of 1847', www.turtlebunbury.
com

Herman Melville, *Moby-Dick*, University of California Press, 1983

Roy L. Behrens, 'Abbott Thayer's Camouflage Demonstrations',
Camoupedia, www.bobolinkbooks.com

Tim Davis, Tim Jones, *The Birds of Lundy*, Devon Bird Watching &
Preservation Society & Lundy Field Society, 2007

John D. Goss-Custard, *The Oystercatcher*, OUP, 1996

Sophia Kingshill, Jennifer Westwood *The Fabled Coast: Legends
and Traditions From Around the Shores of Britain and Ireland*,
Random House, 2012

Solent Forum Nature Conservation Group, 'The Solent Waders
and Brent Goose Strategy', Hampshire and Isle of Wight Wildlife
Trust, November 2010

Caspar Henderson, *The Book of Barely Imagined Beings: A 21st-Century Bestiary*, Granta, 2012

Miles Taylor (ed.), *Southampton: Gateway to the British Empire*, I.B. Tauris, 2007

Horatio Clare, 'South China Sea', *From Our Own Correspondent*, BBC World Service, 4 November 2011

Ken Collins, Jenny Mallinson, 'Solent marine aliens', Report to Solent Forum Nature Conservation Group, January 2011

'*Tireless* Pays Five Day Visit to Southampton', 2 March 2012, www.royalnavy.mod.uk

Olaus Magnus, *Historia de Gentibus Septentrionalibus*, Hakluyt Society, 1998

Tag Barnes, *Waterside Companions*, Arco Books, 1963

Eric Edwards, 'A Fisherman's Lucky Stone', 'England: The Other Within: Analysing the English Collections at the Pitt-Rivers Museum', www.prm.ox.ac.uk

Henry Colley March, 'Witched Fishing Boats in Dorset', *Somerset & Dorset Notes & Queries*, Vol X, 1906

Iris Murdoch, *The Sea, The Sea*, Chatto & Windus, 1978

Callum Roberts, *Ocean of Life*, Allen Lane, 2012

Elaine Morgan, *The Aquatic Ape Hypothesis*, Souvenir Press, 1997

Sir David Attenborough, *Scars of Evolution*, BBC Radio 4, 12 April 2005

Marc Verhagen, Stephen Munro, Mario Vaneechotte, Nicole Oser, Renato Bender, 'The Original Econiche of the Genus *Homo*: Open Plain or Waterside?', *Ecology Research Progress*, Sebastian I. Munoz, Nova Science Publishing, 2007

Greg Downey, 'Human (amphibious model) living in and on the water', *Neuropathology: Understanding the Encultured Brain and Body*, blogs.plos.org

The Asiatic Journal and Monthly Register for British India & its Dependencies, January–July1827

Galveston Daily News, 16 January 1921

Angela Barrett to the author; Jill Gervaise to William Nind

H.W. Brands, *Age of Gold*, William Heinemann, 2005

2. The white sea

'History of the Isle of Wight', Wikipedia.org

Walter de la Mare, *Desert Islands*, Faber & Faber, 1930; Paul Dry Books, 2011

Ward Lock's Guide to the Isle of Wight, Ward Lock, 1950

William Davenport Adams, *The History, Topography, and Antiquities of the Isle of Wight*, Smith, Elder, 1856

Richard Grogan, *Island Life*, Issue 5, Aug/Sept 2006

Oliver Haldane Frazer, 'The History of Marine Mammals off the Isle of Wight', *Proceedings of the Isle of Wight Natural History and Archaeology Society*, Vol 9, 1989

The London Magazine, Or, The Monthly Intelligencer, Vol 27, 8 October 1758

Barbara Jones, *The Isle of Wight*, Penguin, 1950

Nicholas Redman, *Whales' Bones of the British Isles*, Redman Publishing, 2004

Colin Ford, *Julia Margaret Cameron*, National Portrait Gallery, 2003

Jean Claude Feray, 'Virginia Woolf's Indian Ancestor', 15 April 1999, archiver.rootsweb.ancestry.com

Julian Cox, Colin Ford, *Julia Margaret Cameron: The Complete Photographs*, Thames & Hudson, 2003

Hallam Tennyson, *Alfred Lord Tennyson, A Memoir*, Macmillan, 1897

Victoria Gill, 'Tiny songbird northern wheatear traverses the world', 15 February 2012, www.bbc.co.uk/nature

Franz Bairlein et al., 'Cross-hemisphere migration of a 25g songbird', 23 February 2012, Royal Society journal, *Biology Letters*, Vol 8, No 4

Walter Johnson (ed.), *The Journals of Gilbert White*, 6 September 1777, Taylor & Francis, Futura, 1982

Gilbert White, The Natural History of Selborne, Ray Society, London 1993

Thomas Bewick, *British Birds*, Beilby & Bewick, 1797

'Ortolan – Bunting' Wikipedia.org

Thomas Bewick, Mary Trimmer, *A Natural History of the Most Remarkable Quadrupeds, Birds, Fishes, Serpents, Reptiles, and Insects*, Whittingham, 1825

Derek Jarman, *Modern Nature*, Vintage, 1992

Lawrence Wilson, 'Alfred Tennyson', *100 Great Nineteenth-Century Lives*, Methuen, 1983

Brian Hinton, *Immortal Faces: Julia Margaret Cameron on the Isle of Wight*, Isle of Wight County Press, 1992

Nathan J. Emery, 'Are Corvids "Feathered Apes"?', S. Watanbe (ed.), *Comparative Analysis of Minds*, Keio University Press, 2003

Bernd Heinrich, *Mind of a Raven*, Ecco, 2002

'Magpies can "recognise reflection"', 26 May 2009, www.news.bbc.co.uk

Lucy G. Cheke, Christopher D. Bird, Nicola S. Clayton, 'Tool-use and instrumental learning in the Eurasian jay (*Garrulus glandarius*)', *Animal Cognition*, 14 (3), 2011

Joanna Pinnock, 'Feathered Apes', BBC Radio 4, 27 March 2012

E.M. Kirkpatrick (ed.), *Chambers 20th Century Dictionary*, Chambers, 1983

Mark Cocker, *Crow Country*, Jonathan Cape, 2007

Payam Nabarz, *The Mysteries of Mithras*, Inner Traditions, 2005

Thomas Merton, *Bread in the Wilderness*, Catholic Book Club, 1953

Helen Waddell, *Beasts and Saints*, Constable & Co., 1934

R.R. Anderson, *Norse Mythology*, S.C. Griggs, 1879

Betty Kirkpatrick, *Brewer's Concise Dictionary of Phrase and Fable*, Helicon, 1993

William Elliot Griffis, *The Pilgrims in Their Three Homes: England, Holland and America*, Kessinger, 2005

Michelle of Heavenfield, 'St Oswald's English Raven', hefenfelt.wordpress.com

Magnus Magnusson, *Lindisfarne: The Cradle Island*, Tempus, 2004

Bede (trans. J.A. Giles), *The Life and Miracles of St Cuthbert*, Dent, 1910

William Herbert, *The History of the Twelve Great Livery Companies of London*, Guildhall Library, 1837

John McManners, *Cuthbert and the Animals*, Gemini, undated

Reverend Monseigneur C. Eyre, *The History of St Cuthbert*, James Burns, 1859

Durham Account Roll, 1380–81, Durham Cathedral Library

Dominic Marner, *St Cuthbert*, British Library, 2000

Peter Ackroyd, *Poe: A Life Cut Short*, Chatto & Windus, 2008

Francis James Childs, *English and Scottish Ballads*, Little, Brown, 1866

Christopher Newall, *Pre-Raphaelite Vision: Truth to Nature*, Tate, 2004

Sourton village display board, Sourton, Devon

Thomas Merton, *The Seven-Storey Mountain*, Sheldon Press, 1973

William H. Shannon, *Thomas Merton's Dark Path*, Farrar, Straus, Giroux, 1987

Thomas Merton, *Contemplation in a World of Action*, Allen & Unwin, 1971

Beth Cioffoletti, 'The Death of Thomas Merton', fatherlouie. blogspot.co.uk

Thomas Merton, *A Vow of Conversation – Journals 1964–1965*, Farrar, Straus, Giroux, 1988

J.A. Baker, *The Peregrine*, New York Review Books, 2005

Robin S. Hosie, *Reader's Digest Guide to British Birds*, Reader's Digest, 2001

W.H. Gardiner, *Gerard Manley Hopkins: Poems and Prose*, Penguin, 1985

C.A. Hall, *A Pocket Book of Birds*, A. & C. Black, 1936

T.H. White, *The Goshawk*, New York Review Books, 2007

Sylvia Townsend Warner, *T.H. White*, Jonathan Cape/Chatto & Windus, 1967

T.H. White, *England Have My Bones*, Collins, 1936

Richard Jefferies, *The Story of My Heart*, Collins, 1883/1933

'T.H. White & Siegfried Sassoon Correspondence on *The Goshawk*', Harry Ransom Center, eupdates.hrc.utexas.edu

Who's Who, A. & C. Black, 1948

'T.H. White in Alderney', *Monitor*, 13 September 1959, www.bbc. co.uk/archive

'Shallowford Days', www.henrywilliamson.co.uk

3. The inland sea

'Sadler's Wells', *Old and New London*, 1878, Centre for Metropolitan History

John Dineley, 'London Dolphinarium', 2010, www.marineanimal-welfare .com

Benjamin Franklin, *Autobiography and Other Writings*, OUP, 1993

'John Hunter' Royal College of Surgeons, www.rcseng.ac.uk

'John Hunter', Wikipedia.org

Romeo Vitelli, 'The Hanged Man', 31 July 2011, *Providentia: A Biased Look at Psychology in the World*, www.drvitelli.typepad.com

Peter Ackroyd, *Blake*, Sinclair-Stevenson, 1995

Gentleman's Magazine Vol 76, John Nichols, 1797

Ole Daniel Enersen, 'John Hunter', www.whonamedit.com

T. Moore, *Life, Letters & Journals of Lord Byron*, John Murray, 1839

The Times, 2 & 10 March 1826

Jan Bondeson, *The Feejee Mermaid and Other Essays*, Cornell University Press, 1999

William Wordsworth, *The Prelude*, Book 7, 1805

Lynda Ellen Stephenson Payn, *With Words and Knives: Learning Medical Dispassion in Early Modern England*, Ashgate, 2007

Stephen Lewis, 'A pathological misunderstanding', *Wellcome History*, 46, 2011

John Hunter, 'Observations on the structure and œconomy of whales', *Philosophical Transactions of the Royal Society*, 1787/1840

Mark Gardiner, 'The Exploitation of Sea-Mammals in Medieval England: Bones and their Social Context', *Archaeology Journal*, Vol 154, 1997

Ian Riddler, 'The Archeology of the Anglo-Saxon Whale', *The Maritime World of the Anglo-Saxons*, ISAS Monographs, New York, 2013

Salisbury Journal, 21 May 1770

Annual Register, or, A View of the History, Politicks, and Literature, of the Year 1798

Hampshire Chronicle, 9 September 1798

E.J. Slijper, *Whales*, Hutchinson, 1962

Klaus Barthelmess, Ingvar Svanberg, 'Linnæus' Whale', www.idehist.uu.se

J.F. Palmer, *The Works of John Hunter*, Longman, 1837

'The Gunter estate', *Survey of London: Kensington Square to Earl's Court*, 1986, www.british-history.ac.uk

Rob Deaville to the author, 31 October 2011 & 7 December 2012

4. The azure sea

'Charles B. Cory', Wikipedia.org

Jim Enticott, David Tipling, *Seabirds of the World*, New Holland, 2002

José P. Granadeiro, Luis R. Monteiro, Robert W. Furness, 'Diet and feeding ecology of Cory's shearwater', *Marine Ecology Progress Series*, Vol 166, 1998

A.R. Martin, 'Feeding association between dolphins and shearwaters around the Azores Islands', *Canadian Journal of Zoology*, 1986, 64: (6)

E.J. Belda, A. Sánchez, 'Seabird mortality on longline fisheries in the western Mediterranean', SEO/Birdlife, Madrid, Spain, *Biological Conservation* 8, 2001,

Daniel D. Roby, Jan R.E. Taylor, Allen R. Place, 'Significance of stomach oil for reproduction in seabirds', *The Auk*, October 1997

Brett Westwood, Stephen Moss, Chris Watson, 'A Guide to Coastal Birds', BBC Radio 4, 29 August 2010

Hal Whitehead, Peter T. Madsen, Shane Gero, et al., 'Sperm Whales', *Journal of the American Cetacean Society*, Spring 2012, Vol 41, No 1

Thomas I. White, *In Defense of Dolphins*, Blackwell, 2007

Andy Coghlan, 'Whales boast the brain cells that "make us human"', *New Scientist*, 27 November 2006

Hal Whitehead, 'Sperm Whales', *Dominion: A Whale Festival*, Peninsula Arts, Plymouth University, 19 February 2011

Kirsten Thomas, C. Scott Baker, Anton van Helden, Selina Patel, Craig Millar, Rochelle Constantine, 'The world's rarest whale', *Current Biology*, Vol 22, No. 21

Patricia Arranz, Natacha Aguilar de Soto, Peter T. Madsen, Alberto Brito, Fernando Bodes, Mark P. Johnson, 'Following a Foraging Fish-Finder: Diel Habitat Use of Blainville's Beaked Whales Revealed by Echolocation', 7 December 2011, *PLoS One* 6 (12)

Patrick Moore, Lewis Dartnell, *The Sky at Night*, BBC 4, 8 November 2011 & 12 January 2012

'Pico Island, Azores', marineconnection.org

Dr Antonio Fernandez et al., 'Gas and Fat Embolic Syndrome Involving a Mass Stranding of Beaked Whales (Family *Ziphiidae*) Exposed to Anthropogenic Sonar Signals', *Veterinary Pathology*, 42, 2005

Andreas Fahlman, P.H. Kvadsheim, et al., 'Estimated tissue and blood N2 levels and risk of decompression sickness in deep-, intermediate- and shallow-diving toothed whales during exposure to naval sonar', *Frontiers in Aquatic Physiology*, 3:125, 2012

Giuseppe Notarbartolo di Sciara, Alexandros Frantzis, Luke Rendell, 'Sperm Whales: Mediterranean Sperm Whale', *Journal of the American Cetacean Society*, op. cit.

John Pierce Wise Sr, Roger Payne, Sandra S. Wise, Carolyne LaCerte, James Wise, Christy Gianios Jr, W. Douglas Thompson, Christopher Perkins, Tongzhang Zheng, Cairong Zhu, Lucille Benedict, Iain Kerr, 'A global assessment of chromium pollution using sperm whales (*Physeter macrocephalus*) as an indicator species', *Chemosphere*, 75, 2009

Roger Payne to the author, Wellington, 14 March 2010

Sandro Mazzariol, et al., 'Sometimes Sperm Whales (Physeter macrocephalus) Cannot Find Their Way Back to the High Seas: A Multidisciplinary Study on a Mass Stranding', *PLoS One*, May 2011, Vol 6, Issue 5

'Crittervision: The world as animals see (and sniff) it', *New Scientist*, 20 August 2011

M. Klinowska, 'Cetacean live strandings relate to geomagnetic disturbance', *Aquatic Mammals* Vol 11 (1), 1985

Simon Woodings, 'A Plausible Physical Cause for Strandings', Bsc thesis, University of Western Australia, 1995

'197 beached pilot whales die', *The Independent*, 22 February 2011

'Rescuers save 22 melon-headed whales', UPI report, 6 March 2011

Karin Hartman to the author, Pico, 8 July 2011

Hal Whitehead to the author, Plymouth, 19 February 2011

Hal Whitehead, 'Sperm Whales: Capture Me', *Journal of the American Cetacean Society*, op. cit.

Hal Whitehead, *Sperm Whales: Social Evolution in the Ocean*, University of Chicago Press, 2005

Hal Whitehead, Ricardo Antunes, Shane Gero, S.N.P. Wong, D. Engelhaupt, Luke Rendell, 'Multilevel societies of female sperm whales (*Physeter macrocephalus*) in the Atlantic and Pacific: why are they so different?' *International Journal of Primatology*, Vol 33, No 5, 2012

5. The sea of serendipity

David K. Caldwell, Melba C. Caldwell, Dale W. Rice, 'Behaviour of the Sperm Whale, *Physeter catodon* L.', Kenneth S. Norris, *Whales, Dolphins, and Porpoises: Proceedings of the First International Symposium on Cetacean Research, Washington, DC, August 1963*, University of California Press/Cambridge University Press, 1966

Asha de Vos, 'The Sri Lankan Blue Whale Project', whalessrilanka. blogspot.co.uk

Asha de Vos to the author, Weligama, 7 February 2011

Pliny, *Natural History* VI, livingheritage.com/taprobane

Yulia V. Ivashchenko, Phillip Clapham, Robert L. Brownell, Jr, 'Soviet illegal whaling: the Devil and the details', *Marine Fisheries Review*, 73 (3), 2011

'Valentine Orilokova, World's Most Beautiful Captain of a Ship', beautifulrus.com

'The Rebirth of Anaïs Nin's Writing Philosophy', anaisninblog. sybluepress.com

D. Graham Burnett, *Sounding the Whale: Science and Cetaceans in the Twentieth Century*, University of Chicago Press, 2012

'Count de Mauny' www.rootsweb.ancestry.com

Christopher Ondaatje, 'Count de Mauny Island', *The Nation*, www.nation.lk

Tim Street-Porter, 'History of the Island', www.taprobaneisland.com

'Count de Mauny', www.rootsweb.ancestry.com

James S. Romm, *The Edges of the Earth in Ancient Thought: Geography, Exploration, and Fiction*, Princeton University Press, 1992

Sewyn Chomet, Joe Duncan, 'Count de Mauny', www.oscholars. com

Christopher Sawyer-Laucanno, *An Invisible Spectator: A Biography of Paul Bowles*, Bloomsbury, 1989

Paul Bowles, 'How to to live on a part-time island,' *Holiday*, March 1957, www.paulbowles.org

Paul Bowles, 'An Island of My Own', www.taprobaneisland.com

Millicent Dillon, *A Little Original Sin: The Life and Work of Jane Bowles*, Virago, 1988

Richard Hill, 'Taprobane', www.robertehill.co.uk/taprobane

Arthur C. Clarke, BBC *Horizon*, 1964, youtube.com
Juan Pablo Gallo-Reynoso, Janitzo Égido-Villarreal, Guadolupe Martínez-Villalba, 'Reaction of fin whales *Balaenoptera physalus* to an earthquake', *Bioacoustics*, 20, 2011

6. The southern sea

Convict records, Archive Office of Tasmania, CON 31/1/33
Margaret Spence, *Hampshire and Australia, 1783–1791: Crime and Transportation*, Hampshire Record Office, 1992
Robert Hughes, *The Fatal Shore*, Pan Books, 1988
'Convict Ships', *The Times*, 23 August 1846
'Convict Ships', www.jenwillets.com
Jorgen Jorgensen, James Francis Hogan, *The Convict King*, University of Sydney pdf
'The Artists on James Cook's Expeditions', Rudiger Joppien, *James Cook and the Exploration of the Pacific*, Thames & Hudson, 2009
Popular Science Monthly, September 1892, Vol 41, No 42
Hadoram Shirihai, Brett Jarrett, *Whales, Dolphins and Seals: A Field Guide to the Marine Mammals of the World*, A. & C. Black, 2006
George Shelvocke, *A Voyage Round the World by Way of the Great South Sea*, Senex, Longman, &c, 1726
Niels C. Rattenbourg, 'Do birds sleep in flight?', *Naturwissenschaften*, 2006, Vol 93, No 9
Gabrielle A. Nevitt, Francesco Bonadonna, 'Sensitivity to dimethyl sulphide suggests a mechanism for olfactory navigation by seabirds' *Biology Letters*, 22 September 2005; 1(3)
John P. Croxall, Stuart M. Butchart, Ben Lascelles, Alison J. Stattersfield, Ben Sullivan, Andy Symes, Phil Taylor, 'Seabird conservation status, threats and priority actions: a global assessment', *Bird Conservation International* (2012) 22
Reverend John West, *History of Tasmania*, H. Dowling, 1850
Uncle Max Dulumumun Harrison, *Singing Up the Whales*, film, Peter McConchie, 2012
Vivienne Rae-Ellis, *Trucanini: Queen or Traitor*, Australian Institute of Aboriginal Studies, 1981
'Truganini', www.brunyisland.com/truganini
N.J.B. Plomley, *Friendly Mission: The Tasmanian Journals and Papers of George Augustus Robinson, 1829–1834*, Tasmanian Historical Research Association, Quintus, 2008

Michael Desmond, 'Black and White History', *Portrait 32, Magazine of Australian & International Portraiture*, June–August 2009

The Saturday Magazine, 16 February 1833, www.jenwilletts.com/isaac_scott_nind

Charles Manning Clark, Hilary Franklin to Angela Barrett, 15 December 1994

'A Royal Lady – Trucaminni', *The Times*, 6 July 1876

Lyndall Ryan, Neil Smith, 'Trugernanner', *Australian Dictionary of Biography*, National Centre of Biography, Australian National University

W.E.L.H. Crowther, 'Crowther, William Lodewyk', *Australian Dictionary of Biography*, National Centre of Biography, Australian National University

Helen MacDonald, 'The Bone Collectors', *New Literatures Review*, 42, October 2004

Helen MacDonald, *Possessing the Dead: The Artful Science of Anatomy*, Melbourne University Press, 2010

Susan Lawrence, *Whalers and Free Men: Life on Tasmania's Colonial Whaling Stations*, Australian Scholarly Publishing, 2006

The Times, 29 May 1869

The Sydney Gazette and New South Wales Advertiser, 21 April 1805

David S. Macmillan, 'Paterson, William', *Australian Dictionary of Biography*, National Centre of Biography, Australian National University

S. McOrist, A.C. Kitchenor, D.L. Obendorf, 'Skin Lesions in Two Preserved Thylacines: *Thylacinus Cynocephalus*', *Australian Mammal Society*, August 1993, Vo1 16, Part I

Barbara Hamilton-Arnold, *Letters and Papers of G.P. Harris 1803–1812*, 1994, *Imagining the Thylacine: From Trap to Laboratory*, University of Tasmania, www.utas.edu.au

Georges Cuvier, *The Animal Kingdom*, 1827, Joseph Milligan, *Remarks upon the Habits of Wombats*, 1853, cited Dr R. Paddle, *The Last Tasmanian Tiger*, Cambridge University Press, 2000

David Bressan, 'The Last Thylacine', Scientific American blog

Cameron R. Campbell, 'The Thylacine Museum', www.naturalworlds.org/thylacine

Appendices I, II, III, www.cites.org

Bernard Heuvlemans, *On the Track of Unknown Animals*, Hart-Davis, 1958

Hobart *Mercury*, 18 August 1961

Thylacinus Cynocephalus, www.iucnredlist.org

Tim Hilton, *John Ruskin*, Nota Bene/Yale, 2002

Keith Williams, 'Thomas Henry Huxley', *Wellcome History*, Issue 49, Spring 2012

Buck Embey, Jane Oehle Embey, *Thylacine Sightings 1970–1990 in Areas of North Eastern Tasmania Adjacent to the Panama Forest*, July 1990/Jan 2001, Oxford Museum of Natural History collection

Dr Stuart Sleightholme to the author, 2 March 2012

Anthony Hoy, 'Eye on the tiger', Dean Howie's Yowie Research, website

7. The wandering sea

Simon Winchester, *Atlantic: A Vast Ocean of a Million Stories*, HarperPress, 2011

World Register of Marine Species, 15 November 2012, www.marinespecies.org

Anthony Alpers, *A Book of Dolphins*, Jonathan Cape, 1960

T.W. Downes, 'Pelorus Jack, Tuhi-rangi', *Journal of the Polynesian Society*, Vol 23, No. 91, 1914

Brian Fagan, *Beyond the Blue Horizon*, Bloomsbury, 2012

Bernd Brunner, *Bears: A Brief History*, Yale, 2007

Anglo-Saxon Verse, 'The Seafarer'

Charles Kingsley, *The Children's Hereward*, Harrap, 1959

Jonathan Raban, *Passage to Juneau: A Sea and its Meanings*, Vintage, 2000

Hilary Stewart, *Looking at Indian Art of the Northwest Coast*, Douglas & McIntyre, 1979

'Whales in Māori tradition' *Te Ara: The Encyclopaedia of New Zealand*, www.teara.govt.nz

Transactions and Proceedings of the Royal Society of New Zealand, Vol 5, 1872

Gray Chapman, 'The Day the Whales Died', www.wainuibeach.co.nz

Witi Ihimaera to the author, 16 January 2012

Witi Ihimaera, 'World Service Book Club', BBC World Service, 8 January 2012

B.J. Marlow and J.E. King, 'Sea Lions and Fur Seals of Australia and New Zealand: The Growth of Knowledge', *Australian Mammal Society*, October 1974

'Fyffe House' www.historicplaces.org.nz

Atholl Anderson, 'On Evidence for Survival of Moa in European Fiordland', *New Zealand Journal of Ecology*, Vol 12, Supplement, 1989

Errol Fuller, *Extinct Birds*, Viking/Rainbird, London, 1987

David Bressan, 'Ka ngaro i te ngaro a te Moa', 2011, blogs. scientificamerican.com

Nicola Brown, 'What the Alligator Didn't Know: Natural Selection and Love in *Our Mutual Friend*', *Interdisciplinary Studies in the Long Nineteenth Century*, No 10, 2010, www.19.bbk.ac.uk

Roy Mackal, *Searching for Hidden Animals*, Doubleday, 1980

'Birdman says moa surviving in the bay', 5 January 2008, www. hawkesbaytoday.co.nz

Daniel Cressy, 'Life thrives in ocean canyon', *Nature*, 27 April 2010

Jon Ablett to the author, Natural History Museum, 2 September 2011

Mike Donoghue to the author, Kaikoura, 18 March 2010

Alastair Judkins to the author, Kaikoura, 18 March 2010

International Union for Conservation of Nature Red List, *Diomedea epomophra*, www.iucnredlist.org

'Sperm whale-watching' review August 2012, www.doc.govt.nz

'Whale-watching in danger', www.royalsociety.org.nz

Kauahi Ngapora to the author, Kaikoura, 9 March 2010

8. The silent sea

Greg Dening, *Islands and Beaches*, University Press of Hawaii, 1980

Elsdon Best, *The Māori as he was: A Brief Account of Life as it was in Pre-European Days*, Dominion Museum, Wellington, 1934

Johannes C. Andersen, 'New Zealand Bird-song; Further Notes', *Transactions and Proceedings of the Royal Society of New Zealand, 1868–1961*, Vol 47

Brenda M. White, 'Traill, Thomas Stewart', *Oxford Dictionary of National Biography*, OUP, 2004

Thomas Traill, *Memoir of William Roscoe*, Smith, Watts, 1853

C.F.A. Marmoy, 'The "Auto-Icon" of Jeremy Bentham at University College, London', Thane Library of Medical Sciences, UCL

Jeremy Bentham, *An Introduction to the Principles of Morals and Legislation*, 1789, Pickering, 1823

William Scoresby, *Account of the Arctic Regions*, The Religious Tract Society, 1851

William Scoresby, *A Journal of a Voyage to the Northern Whale-Fishery*, Constable, 1823

Journal of Natural Philosophy, Chemistry, and the Arts, February 1809

George Lillie Craik, *The New Zealanders*, Society for the Diffusion of Useful Knowledge, 1830

John Savage, *Some Account of New Zealand*, 1807, Hocken Library, 1966

Stephen Oliver, 'Te Pehi Kupe', *Dictionary of New Zealand Biography, Te Ara: The Encyclopaedia of New Zealand*

Geoffrey Sanborn, 'Whence Come You, Queequeg?' *American Literature*, Vol 77, No 2, June 2005, Duke University Press

Peter B. Maling, 'Langlois, Jean-François', *Dictionary of New Zealand Biography, Te Ara: The Encyclopaedia of New Zealand*

Derex Cox to the author, Akaroa, 16 March 2010

Dr Barbara Maas, 'The Catch with New Zealand's Dolphins', World Whale Conference speech, Brighton, 26 October 2012

Alan N. Baker, Adam N.H. Smith, Franz B. Pichler, 'Geographical variation in Hector's dolphin: recognition of new subspecies of *Cephalorhynchus hectori*', *Journal of the Royal Society of New Zealand*, Vol 32, No 4

9. The sea in me

Richard Jefferies, *The Story of My Heart*, 1883

David Rothenberg, *Thousand Mile Song*, Basic Books, 2008

Brewer's, op. cit.; Magnus, op. cit.

Gilbert White, Letter XIII, *Natural History of Selborne*, T. Bensley, 1789

Ted Hughes, 'Work and Play', *Collected Works*, Faber, 2005

Steve Connor, 'Pole to pole: the extraordinary migration of the arctic tern', *The Independent*, 12 January 2010

Index

Page numbers in *italic* refer to illustrations

361

368